Natural Gas Hydrates—Energy Resource Potential and Associated Geologic Hazards

Edited By
T. Collett, A. Johnson, C. Knapp, and R. Boswell

AAPG Memoir 89

Co-published by

The American Association of Petroleum Geologists
Tulsa, Oklahoma
and
U.S. Department of Energy, National Energy Technology Laboratory (NETL)
and
AAPG Foundation
and
AAPG Energy Minerals Division

AAPG Editor: Gretchen M. Gillis
AAPG Geoscience Director: James B. Blankenship

ON THE COVER: Burning synthetic (man-made) methane hydrate; photo courtesy of the U.S. Department of Energy.

This publication is available from:

The AAPG Bookstore
P.O. Box 979
Tulsa, OK U.S.A. 74101-0979
Phone: 1-918-584-2555 or 1-800-364-AAPG (U.S.A. only)
E-mail: bookstore@aapg.org
www.aapg.org

Canadian Society of Petroleum Geologists
600, 640 – 8th Avenue S.W.
Calgary, Alberta T2P 1G7
Canada
Phone: 1-403-264-5610
Fax: 1-403-264-5898
E-mail: reception@cspg.org
www.cspg.org

Geological Society Publishing House
Unit 7, Brassmill Enterprise Centre
Brassmill Lane, Bath BA13JN
United Kingdom
Phone: +44-1225-445046
Fax: +44-1225-442836
E-mail: sales@geolsoc.org.uk
www.geolsoc.org.uk

Affiliated East-West Press Private Ltd.
G-1/16 Ansari Road, Darya Gaaj
New Delhi 110-002
India
Phone: +91-11-23279113
Fax: +91-11-23260538
E-mail: affiliate@vsnl.com

Contribution Acknowledgments

AAPG wishes to thank the following for their generous contributions to

Natural Gas Hydrates—Energy Resource Potential and Associated Geologic Hazards

U.S. Department of Energy, National Energy Technology Laboratory (NETL)

AAPG Foundation

AAPG Energy Minerals Division

Contributions are applied toward the production cost of the publication, thus directly reducing the book's purchase price and making the volume available to a larger readership.

About the Editors

Dr. Timothy S. Collett is a research geologist in the Geologic Division of the U.S. Geological Survey (USGS). Most recently, Collett was a co-chief scientist and operational manager for the India National Gas Hydrate Program Expedition-01 gas-hydrate research project. Collett was a co-chief scientist of the international cooperative research project responsible for drilling gas-hydrate production research wells in the Mackenzie delta of Canada under the Mallik 1998 and 2002 efforts. Collett sailed as the logging scientist on the Ocean Drilling Program (ODP) Legs 164 and 204 gas hydrate research cruises. Collett was also the logging scientist on the Gulf of Mexico Joint Industry Project Gas Hydrate Research Cruise in 2005 and a Co-Chief Scientists on the Integrated Ocean Drilling Program (IODP) Expedition 311. Collett was the principal investigator responsible for organizing and conducting the 1995 USGS National Oil and Gas Assessment of natural gas hydrates. Collett holds a B.S. in geology from Michigan State University, a M.S. in geology from the University of Alaska, and a Ph.D. from the Colorado School of Mines.

Art Johnson is president and chief of exploration for Hydrate Energy International (HEI) in Kenner, Louisiana and is engaged in exploration efforts throughout the world. Prior to forming HEI in 2002, Art was with Chevron for 25 years. He is co-chair of the AAPG/Energy Minerals Division Gas Hydrate Committee and is a past-president of New Orleans Geological Society. Art chaired the federal Methane Hydrate Advisory Committee from 2001 to 2006 and has advised the U.S. Congress and the White House on energy issues since 1997. He has an ongoing role coordinating the research efforts of industry, universities, and government agencies, and serves as an AAPG visiting geoscientist.

Camelia C. Knapp received her Ph.D. in geophysics from Cornell University (2000) and her B.S. and M.S. degrees in geophysical engineering from the University of Bucharest in Romania (1988). She worked with the Romanian State Oil Company and the National Institute for Earth Physics in Romania for several years. She was a Fulbright fellow at Cornell University before pursuing her Ph.D. Currently, an assistant professor in the Department of Geological Sciences at the University of South Carolina, her research interests include exploration and environmental geophysics, crustal-scale seismology, and gas hydrates. She is also the director of undergraduate studies.

Ray Boswell manages methane hydrate research and development programs at the U.S. Department of Energy (DOE) National Energy Technology Laboratory in Morgantown, West Virginia. He currently chairs the Interagency Technical Coordination Team for Gas Hydrates that includes representatives of seven federal agencies, and has participated in the planning and operations of gas-hydrate field programs in India, Alaska, and the Gulf of Mexico. Previously, he managed tight gas sands research and development programs for DOE and conducted geologic-based resource assessments for tight gas resources in Appalachian, mid-continent, and the Rocky Mountain basins. Ray holds a Ph.D. in geology from West Virginia University.

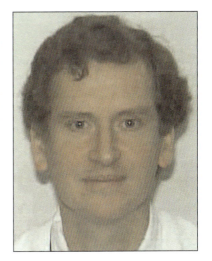

Acknowledgements

The editors of this Memoir want to first acknowledge that this effort began in September, 2004 with the American Association of Petroleum Geologists (AAPG) sponsorship of the Hedberg Research Conference on *"Natural Gas Hydrates: Energy Resource Potential and Associated Geologic Hazards"* in Vancouver, British Columbia, Canada. The success of this conference can be largely credited to the organizational skills of Debbi Boonstra with the AAPG Education Department. Financial support for this conference was provided by the Gulf of Mexico Gas Hydrates Joint Industry Project, Schlumberger, and the U.S. Department of Energy. The Energy Minerals Division of the AAPG was also a co-sponsor of the conference, keeping with their long history of support for gas-hydrate research and development issues. The Canadian Association of Petroleum Geologists and the Geological Survey of Canada also contributed to the organization of the conference.

We also want to thank the AAPG Publications Committee for accepting our proposal for this Memoir on gas hydrates. We also greatly appreciate all of the support and guidance provided to this effort by Beverly Molyneux and the AAPG Technical Publications group. Special thanks are extended to Frances Whitehurst as the lead AAPG production manager for this effort. The publication of this Memoir was also supported by financial contributions from the U.S. Department of Energy and by an award from the AAPG Foundation, as arranged by the Energy Minerals Division of the AAPG.

As the editors of this volume, we want to extend our greatest appreciation to the more than 150 authors who contributed their research efforts to this Memoir. Both your intellectual contributions and support for this effort have been inspirational. We also need to acknowledge our research colleagues who provided our authors with more than 100 individual technical reviews—their support, suggestions, and always constructive criticism was greatly appreciated.

At a personal level, we also need to acknowledge the support of our own employers and companies that supported our editorial contributions, including the Energy Resources Program of the U.S. Geological Survey, Hydrate Energy International, the University of South Carolina, and the U.S. Department of Energy—National Energy Technology Laboratory.

Foreword

In September, 2004 the American Association of Petroleum Geologists (AAPG) convened a Hedberg Research Conference in Vancouver, British Columbia, Canada titled *"Natural Gas Hydrates: Energy Resource Potential and Associated Geologic Hazards"*. The primary goals of the conference were to critically examine the geologic parameters that control the occurrence and stability of gas hydrates, assess the volume of gas stored within gas-hydrate accumulations, assess exploration methods for identifying commercial-gas hydrate prospects, identify the technologies needed to economically produce gas from hydrate, assess possible marine-slope-stability hazards that can be attributed to the occurrence of gas hydrate, and analyze the effects of gas hydrate on drilling safety. The conference was attended by 120 invited participants, coming from more than 13 different countries, reflecting the multi-national interest in gas hydrates. The 75 U.S. participants included 29 from government agencies, 26 from academia, and 20 from industry. The 47 participants from outside the U.S. had similar diverse affiliations. The conference featured 43 oral presentations, 46 poster presentations, three formal discussion sessions, and a key-note address by Marlan Downey titled: *Boulders in the Path, Problems on the Way Towards The Gas Hydrate Rainbow*. The conference concluded with a panel discussion on geology and energy resource potential of gas hydrates.

In the discussion and panel sessions, various participants expressed a sense that significant progress was being made in addressing some of the key issues on the formation, occurrence, and stability of gas hydrates in nature. The concept of gas-hydrate petroleum systems, as they compare to conventional oil and gas petroleum systems, was gaining acceptance. In fact, the use of complex numerical modeling is now allowing the components of gas-hydrate petroleum systems (i.e., source, migration, trap, and timing) to be assessed and quantified. However, there was a growing appreciation that some of the processes leading to the formation of gas hydrates in marine versus permafrost environments may differ. For marine gas-hydrate exploration, several groups expressed the growing need for better geophysical methods that would allow for the direct detection and evaluation of gas-hydrate accumulations; it was concluded the era of assessing marine gas hydrates through the mapping of bottom simulating seismic reflectors (BSRs) alone is drawing to a close.

As reviewed during the conference, several studies from the early 2000s estimate the volume of gas trapped in global gas-hydrate accumulations to be significantly less than some widely cited early estimates. However, by all estimates, the volume of gas trapped in hydrates likely exceeds that trapped in all currently-producible gas accumulations. It was highlighted that none of the existing assessments have predicted how much gas can be produced from the world's gas-hydrate accumulations, and it was concluded that much more work is needed to go beyond the existing in-place gas-hydrate volumetric estimates. Industry representatives also expressed the importance of relatively small, well-defined gas-hydrate accumulations in the one to five trillion cubic feet (~0.03-0.14 trillion cubic meters) range that can be drilled, tested, and possibly produced. Marlan Downey challenged the audience to "think bigger" and view gas-hydrate production as a potential paradigm shift of global importance.

Numerous presenters reported on the results of the Mallik 2002 Gas Hydrate Production Research Well Program. The Mallik 2002 gas-hydrate production testing and modeling effort has for the first time allowed the rational assessment of the production response of a gas hydrate accumulation. The conference participants concluded that a key goal of the industry contribution should be to document that commercial rates of gas production from hydrates are possible. It is important to note that two independent studies of the deliverability of gas from gas hydrates on the North Slope of Alaska and the Canadian Arctic reported similar economics, and a gas delivery cost of about $4-5 (USD) per thousand cubic feet of gas (~28 cubic meters of gas) (2004). Several participants noted that gas-hydrate production may require government incentives, much like the early days of coalbed methane production in the U.S. But probably the most significant conclusion coming from the discussions in Vancouver was the strong statement that we need more dedicated and expanded field production testing to assess the ultimate resource potential of gas hydrates.

Some of the liveliest discussions during the conference focused on the potential hazards associated with gas hydrate especially slope-stability issues (both natural and human induced). One of the major conclusions of these discussions was the acknowledgment that more effort is needed to document case histories of actual gas-hydrate-induced drilling and completion problems.

For more information on the Hedberg Research Conference on gas hydrates please see the "Education" section of the AAPG web site <http://www.aapg.org>. All of the extended abstracts published in the conference proceedings volume are posted on the AAPG web site.

As a continuation of the Hedberg Research Conference in Vancouver, the conveners of the conference and the editors of this Memoir have worked with more than 150 authors and coauthors to prepare this Memoir on gas hydrates. This publication follows the goals of the Hedberg conference; however, the contents of this Memoir were expanded to include all aspects of gas hydrates in nature. This Memoir contains 39 individual contributions, ranging from long, topical summaries to shorter, focused research papers. This Memoir has been published in two parts, with digital versions of all the complete research papers included on the enclosed DVD. The hardcopy portion of the Memoir includes abstracts and several key figures for each of the contributions along with a complete copy of a gas hydrate technical review.

The digital portion of this Memoir has been organized into a series of topical sections, with the first contribution (Chapter 1) being a gas hydrate technical review in which the Memoir editors have provided a comprehensive examination of the components of a gas-hydrate petroleum system, analysis of ongoing and recently completed international gas-hydrate geologic and resource studies in both marine and arctic permafrost environments, and the examination of gas-hydrate production studies and modeling efforts.

The next section of the digital portion of this Memoir contains two additional review articles (Chapters 2–3) that describe the accomplishments of the national gas-hydrate research programs in the United States and Japan. The marine gas-hydrate section contains 17 contributions (Chapters 4–20), with one of the highlights being the inclusion of the research results from the Japan National Gas Hydrate Program. Also included in the marine gas-hydrate section of the Memoir are several contributions dealing with the geology and geophysical properties of gas-hydrate occurrences in the Gulf of Mexico, the Bering Sea and along the eastern and western continental margins of North America. Other contributions deal with gas-hydrates off the western coast of Norway, and offshore New Zealand and Taiwan. The Arctic terrestrial gas-hydrate contributions include nine reports, all dealing with gas-hydrate occurrences on either the North Slope of Alaska or in the Mackenzie River Delta of Canada. Five of the Arctic related contributions (Chapters 21–29) summarize the results of the 2003–2004 gas-hydrate drilling effort in northern Alaska lead by Anadarko.

The last two sections of this Memoir deal with gas-hydrate laboratory and modeling studies. The eight laboratory contributions (Chapters 30–37) focus on the analysis of gas-hydrate samples either recovered from nature or made in the laboratory. The measurements discussed in the laboratory contributions range from characterizing the nature of gas-hydrate occurrences in sediments to numerous physical property measurements, including thermal properties, permeability of gas-water-hydrate systems, acoustic and mechanical strength properties, along with the characterization of thermodynamic and kinetic controls on gas-hydrate stability and dissociation. The two hydrate modeling studies (Chapters 38–39) also examined the kinetics of hydrate phase transitions and hydrate formation in marine sediments.

The review and editorial process for this Memoir have followed the guidelines as specified by the AAPG, with the completion of at least two technical reviews for each contribution and additional editorial reviews as required. Each paper was also reviewed by the AAPG for appropriate grammar and formatting. It is also acknowledged, that the Hedberg Research Conference on gas hydrates was held four years ago. Also, the first manuscripts for this publication were prepared and submitted more than three years ago. Both issues raise the question of the timeliness of this publication. The editors have worked with the authors to identify outdated material and update each manuscript as needed. We believe the editorial standards we have adopted and the work of the contributing authors to update their manuscripts has addressed any concerns associated with the timeliness of this publication. We also acknowledge that the average length of the manuscripts included in this publication is in some cases longer than most AAPG reports. Within this Memoir, however, we have had the unique opportunity to include comprehensive reviews of important gas hydrate research projects that would not be easily published elsewhere.

Because of the rapidly emerging worldwide interest in gas hydrates, we believe this comprehensive treatise on the geology of gas hydrates will be well received by both the gas-hydrate research community and exploration/development geologists working in arctic and deep marine environments. As highlighted, this publication has mostly followed the resource assessment theme of the 2004 Hedberg Research Conference on gas hydrates along with other significant additions on gas hydrate related geologic hazards, which are also of interest to the energy industry active in deep marine and arctic environments.

Note to the Reader

This publication is a continuation of AAPG's efforts to combine the best of both digital and print publications for its members. Memoir 89, Natural Gas Hydrates—Energy Resource Potential and Associated Geologic Hazards, consists of two sections.

The Extended Abstracts for 39 papers are printed in their entirety in this volume. The CD-ROM located in the inside back cover of this book contains the full 39 papers. Both the Extended Abstracts and the full papers are in full color.

The dual digital and print format allows authors greater scope for both length of manuscripts and number of color figures, and provides the reader much more versatility in using the content. For example, figures can be used directly for making slides or Microsoft PowerPoint© presentations. Individual papers can easily be printed for later reading, and key word searching through the papers is possible.

Finally, this format allows AAPG the flexibility to publish a volume that contains material that will appeal to a wider audience with additional interests.

Table of Contents

Allison, E. C., and R. M. Boswell, 2009, Extended abstract—Overview of the United States Department of Energy's gas-hydrate research program: 2000 to 2005, *in* T. Collett, A. Johnson, C. Knapp, and R. Boswell, eds., Natural gas hydrates—Energy resource potential and associated geologic hazards: AAPG Memoir 89, p. 5–7.

Extended Abstract—Overview of the United States Department of Energy's Gas-hydrate Research Program: 2000 to 2005

Edith C. Allison

U.S. Department of Energy, Office of Fossil Energy, Washington, DC, U.S.A.

Ray M. Boswell

U.S. Department of Energy, National Energy Technology Laboratory, Morgantown, West Virginia, U.S.A.

ABSTRACT

Gas hydrate has been a target of research by the United States Department of Energy (DOE) for more than two decades. Since 2000, an accelerated DOE research and development effort has included efforts to improve the understanding of gas-hydrate occurrence, its behavior under dynamic conditions, and its potential as a future energy source. The DOE has supported several important accomplishments: improved the understanding of the fundamental physical and chemical properties of gas hydrate and gas-hydrate-bearing sediments; significant strides in understanding how to detect and characterize gas-hydrate accumulations; improved understanding of the complexity of gas hydrate in nature; development of new tools to sample, measure, and monitor gas hydrate in the field; and the development of the first reservoir models of gas hydrates.

Ongoing work is expanding and extending these accomplishments within five broad categories: laboratory studies, modeling, exploration technologies, field studies, and field sample collection and analysis tool development. This work is being conducted through several cooperative agreements with universities and industries through funding for specific activities within the DOE National Laboratory system and through interagency agreements with the U.S. Geological Survey and Naval Research Laboratory.

FIGURE 1. The science crew of the Chevron-led Joint Industry Project's 2005 expedition describe sediments obtained from the Atwater Valley site in the Gulf of Mexico.

FIGURE 3. The integrated pressure testing chamber (IPTC) is one of the advanced tools for marine gas-hydrate investigations developed within the Department-of-Energy-led program.

FIGURE 2. A sample of gas-hydrate-bearing sandstone obtained during the Department of Energy-BP-U.S. Geological Survey 2007 field program at Milne Point, Alaska.

LOOKING FORWARD

Over the next 5 years, the program will continue to focus its efforts on the research and goals stated in the MHR&D Act. Specifically, we will (1) work with our industry and academic partners to conduct and analyze long-term production tests of a gas hydrate reservoir, (2) refine the methodologies for reliable marine gas hydrate detection through a variety of remote sensing technologies, (3) develop reliable pressure coring tools and pressurized sample analysis equipment, and (4) integrate the growing knowledge of gas hydrate occurrence and behavior into the global carbon and climate change models to more fully understand the potential and impacts of methane release from gas hydrate, both from natural and man-made causes. To support these efforts, the DOE will continue to support focused laboratory studies to provide improved input parameters for the program's numerical models and will continuously test and improve those models as more data become available. In addition, the DOE will support the creation of an internationally accessible and comprehensive database of gas hydrate data. In pursuing this work, the DOE will continue its strong record of (1) providing significant educational opportunities for the next generation of energy scientists, (2) timely publication of R&D results, and (3) leveraging federal dollars and advancing gas hydrate science through collaboration with other nations. Finally, the program will continue to strive to implement the most salient recommendations of the NRC (2004), including the continued and expanded use of external merit reviews in the process of selecting, monitoring, and evaluating the work done under the program.

ACKNOWLEDGMENTS

The authors wish to acknowledge all of DOE's research partners as well as the members of the collaborating federal agencies and the federal advisory committee for their continued contributions to excellence in the federal gas hydrate R&D program.

Tsuji, Y., T. Namikawa, T. Fujii, M. Hayashi, R. Kitamura, M. Nakamizu, K. Ohbi, T. Saeki, K. Yamamoto, T. Inamori, N. Oikawa, S. Shimizu, M. Kawasaki, S. Nagakubo, J. Matsushima, K. Ochiai, and T. Okui, 2009, Extended abstract—Methane-hydrate occurrence and distribution in the eastern Nankai Trough, Japan: Findings of the Tokai-oki to Kumano-Nada methane-hydrate drilling program, *in* T. Collett, A. Johnson, C. Knapp, and R. Boswell, eds., Natural gas hydrates—Energy resource potential and associated geologic hazards: AAPG Memoir 89, p. 9–12.

3

Extended Abstract—Methane-hydrate Occurrence and Distribution in the Eastern Nankai Trough, Japan: Findings of the Tokai-oki to Kumano-nada Methane-hydrate Drilling Program

Yoshihiro Tsuji, Tetsuya Fujii, Masao Hayashi, Ryuta Kitamura, Masaru Nakamizu, Katsuhiro Ohbi, Tatsuo Saeki, and Koji Yamamoto
Japan Oil, Gas and Metals National Corporation (JOGMEC), Chiba, Japan

Takatoshi Namikawa, Takao Inamori, Nobutaka Oikawa, and Shoshiro Shimizu
Japan Petroleum Exploration Co., Ltd., Tokyo, Japan

Masayuki Kawasaki and Sadao Nagakubo
Japan Drilling Co., Ltd., Tokyo, Japan

Jun Matsushima
The University of Tokyo, Tokyo, Japan

Koji Ochiai
Inpex Corp., Tokyo, Japan

Toshiharu Okui
Tokyo Gas Co., Ltd., Tokyo, Japan

ABSTRACT

The widespread distribution of bottom-simulating reflectors (BSRs) is known in the Nankai Trough area off the southern coast of central Japan from seismic data collected for oil and gas exploration purposes. Japan has been investigating the occurrence and resource potential of methane hydrate in the area of the BSR-inferred gas-hydrate occurrences in the Nankai Trough since the mid-1990s.

After obtaining the result of the Ministry of International Trade and Industry (MITI) Nankai Trough exploratory drilling program in 1999 and 2000 and finding methane hydrate in turbidite sands, more drilling and seismic studies of hydrate distribution and resource-volume assessments were recommended and conducted.

Subsequent to the MITI Nankai Trough drilling program, widely spaced two-dimensional (2-D) and selected high-resolution three-dimensional (3-D) seismic surveys were conducted in the Nankai Trough area from offshore Tokai to the Kii Peninsula in 2001 and 2003. The BSR distribution was further examined in the newly acquired seismic data, and a multiwell drilling program consisting of coring, logging while drilling (LWD), and wireline logging followed in early 2004.

For the 2004 drilling program, 16 new drill sites at water depths ranging from 720 to 2033 m (2362 to 6670 ft) were selected based on a seismic analysis of stratigraphic features and the nature of BSRs. Some locations showed clear BSRs with variable reflection amplitudes, some had vague BSRs between locations that had clear BSRs, and some exhibited double BSRs. Also, some of the locations were characterized by seismically inferred high-velocity zones with and without BSRs.

Drilling operations, conducted from the drill ship, *JOIDES Resolution*, included (1) first drilling dedicated LWD holes to obtain logging data for resource assessments to 100 m (328 ft) below the BSRs, (2) wireline logging to calibrate the LWD response at two well sites, which were selected using the result of the LWD data, (3) continuous coring to calibrate the wireline data at two wireline-logged well sites, and (4) spot-coring wells to collect methane-hydrate samples at well sites where methane-hydrate occurrence was inferred in the LWD and wireline log data.

Upon drilling, hydrate-bearing sediments were easily identified by the comparison of resistivity and density log data from both LWD and wireline surveys. Methane-hydrate-bearing sediments were confirmed at most of the sites where BSRs were identified, although the thicknesses of the actual hydrate occurrences varied. At some of the locations, hydrate-saturated turbidite sands, ranging in thickness up to about 1 m (3 ft) or less, were found interbedded with non-hydrate-bearing, mud-dominated sediments. In some cases, the interbedded hydrate-bearing section or hydrate-concentrated zones reached local total thickness of more than 105 m (344 ft). Also, in other locations, methane-hydrate-bearing silty sediments and nodular or fracture-filling massive methane hydrate in muddy sediments were observed. Hydrate-saturated turbidite sands, hydrate-bearing silty sediments, and massive methane-hydrate samples in muddy sediments were collected by pressure coring with the pressure temperature coring system (PTCS) and/or with conventional wireline coring systems.

Although most of the methane-hydrate-bearing sediments identified in the wells were associated with BSRs, some of the hydrate-saturated turbidite sands that occurred well above the base of a gas-hydrate stability zone did not have obvious BSRs below them. Wells without well-developed hydrate-bearing sediments were also drilled at locations with well-developed BSRs. In some cases, hydrate-concentrated sediments are clearly shown as high-velocity zones by seismic velocity analysis.

It has been shown that BSRs are not always good indicators of thick concentrated methane-hydrate occurrences and cannot be used to accurately predict in-situ methane-hydrate volumes, although a BSR is an important indicator of methane hydrate.

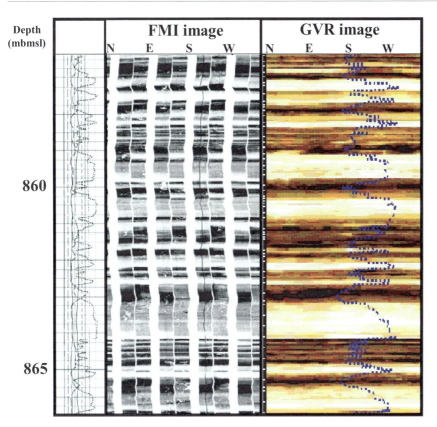

FIGURE 1. Comparison of a wellbore image between wireline full bore formation microimager (FMI) and logging-while drilling GeoVision Resistivity (LWD GVR) in a methane-hydrate-concentrated zone. In both logging images, the light-colored part is high-resistivity and methane-hydrate-filled sand, and the dark-colored part is low-resistivity and fine sediments with no methane-hydrate filling. Depth is shown as meters below mean sea level (mbmsl). 1 m = 3.3 ft; FMI and GVR are marks of Schlumberger.

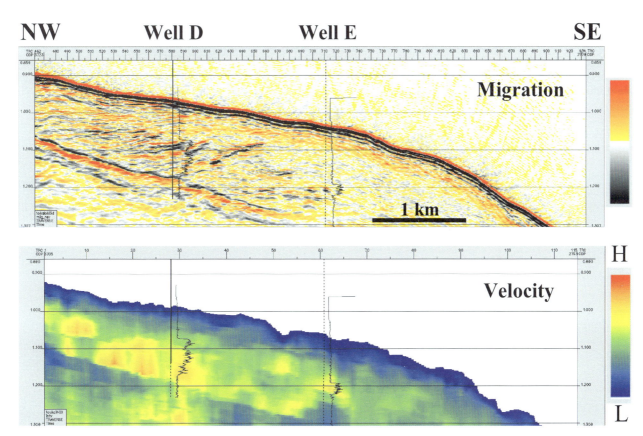

FIGURE 2. Methane-hydrate-concentrated zone with relation to velocity distribution. The top is a migrated section and the bottom is the velocity analysis result. The log curves show compressional wave traveltime. Well D has a thick high-velocity zone to BSR depth, and well E has a thin high-velocity zone and no BSR below. Depths are shown in ms for the seismic sections. H and L show higher or lower velocities respectively.

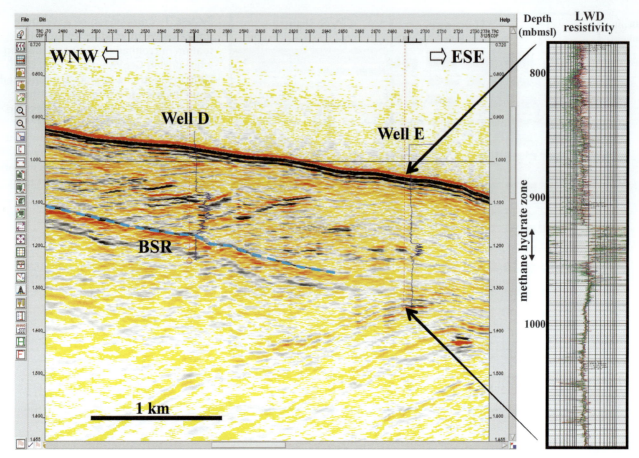

FIGURE 3. A highly concentrated zone of methane-hydrate-bearing sands confirmed separately above the base of the gas-hydrate stability zone (GHSZ) at well E of Figure 2. The bottom-simulating reflector (BSR) is not evident at the well locality. Depths are shown in msec for the seismic section and meters below mean sea level (mbmsl) for the logging-while-drilling (LWD) resistivity (in ohm m).

The multiwell drilling program, entitled Ministry of Economy, Trade and Industry (METI) Tokai-oki to Kumano-nada, provided important results and information on the occurrence of methane hydrate in the Nankai Trough and its relationship to the presence of BSRs. These results, combined with the results of ongoing seismic, sedimentological, and geochemical studies, will continue to be applied in the understanding of the methane-hydrate systems and in the assessment of methane-hydrate resources offshore Japan.

Hardage, B. A., P. E. Murray, R. Remington, M. De Angelo, D. Sava,
H. H. Roberts, W. Shedd, and J. Hunt Jr., 2009, Extended abstract—
Multicomponent seismic technology assessment of fluid-gas expulsion
geology and gas-hydrate systems: Gulf of Mexico, *in* T. Collett, A. Johnson,
C. Knapp, and R. Boswell, eds., Natural gas hydrates—Energy resource
potential and associated geologic hazards: AAPG Memoir 89, p. 13–15.

4

Extended Abstract—Multicomponent Seismic Technology Assessment of Fluid-gas Expulsion Geology and Gas-hydrate Systems: Gulf of Mexico

B. A. Hardage, P. E. Murray, R. Remington, M. De Angelo, and D. Sava
Bureau of Economic Geology, Austin, Texas U.S.A.

H. H. Roberts
Louisiana State University, Baton Rouge, Louisiana, U.S.A.

W. Shedd and J. Hunt Jr.
Minerals Management Service, New Orleans, Louisiana, U.S.A.

ABSTRACT

Four-component ocean-bottom-cable (4-C OBC) seismic data acquired in deep water across the Gulf of Mexico were used to study near-sea-floor geologic characteristics of fluid-gas expulsion systems. These expulsion features extended from great depths to the sea floor and were ideal conduits by which deep thermogenic gases could migrate upward into the hydrate stability zone. Although the 4-C OBC data used in this study were acquired to evaluate oil and gas prospects far below the sea floor, we show that the data have great value for studying near-sea-floor geology. Research results amassed in this study stress the importance of the converted-shear-wave (P-SV) mode extracted from 4-C OBC data for studying strata shallower than the base of the hydrate stability zone. In deep water, the P-SV mode creates an image of near-sea-floor strata that has a spatial resolution an order of magnitude better than the resolution of compressional (P-P) seismic data, regardless of whether the compressional wave data are acquired with OBC technology or with conventional towed-cable seismic technology. This increased resolution allows the P-SV mode to define seismic sequences, seismic facies, small-throw faults, and small-scale structures in the interval between the sea floor and the base of the hydrate stability zone that cannot be detected with compressional wave seismic data.

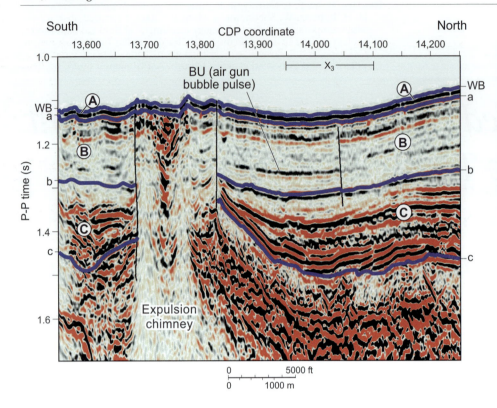

FIGURE 1. Ocean-bottom-cable (OBC) compressional wave (P-P) image along Line 549, Block GC204. WB = water bottom. Horizons a–c are interpreted chronostratigraphic boundaries. Intervals A–C are interpreted chronostratigraphic units. Vertical solid lines are interpreted faults. BU is an air gun bubble pulse, not a geologic boundary. Compressional wave velocity (V_p), shear wave velocity (V_s), and V_p/V_s analyses were done across interval X_3. The horizontal axis is defined in terms of CDP coordinates, where CDP = common depth point.

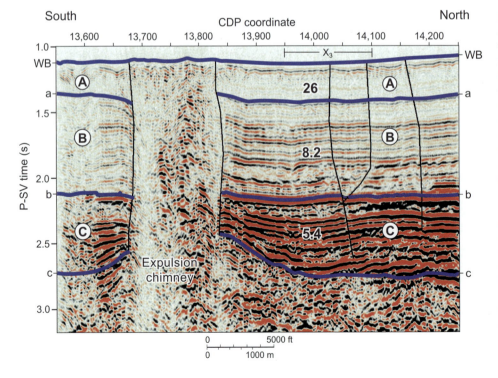

FIGURE 2. Ocean-bottom-cable (OBC) converted shear wave (P-SV) image along Line 549, Block GC204. Labeled horizons and units are interpreted to be depth equivalent to the same labeled features on the compressional wave (P-P) image of Figure 1. Water-bottom (WB) horizon from the P-P images is transferred onto this image for reference. Vertical solid lines are interpreted faults. The number labeled on each unit is the average of seismic-based estimates of compressional wave velocity/shear wave velocity for that unit over the interval labeled X_3. The horizontal axis is defined in terms of CDP coordinates, where CDP = common depth point.

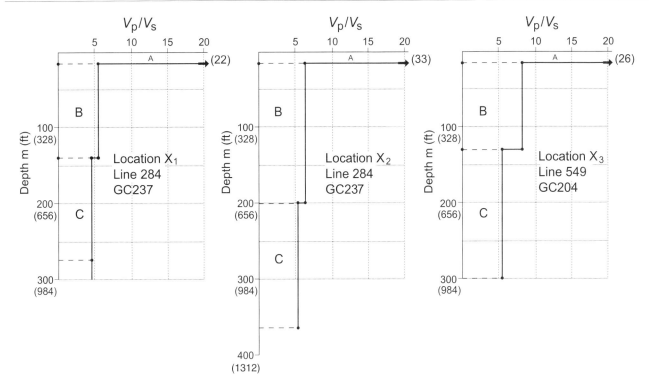

FIGURE 3. Depth-based profiles of averaged compressional wave velocity/shear wave velocity (V_p/V_s) ratios across intervals X_1 and X_2 of Line 284 (Chapter 4 on the CD in the back of this book) and X_3 of Line 549 (Figure 2).

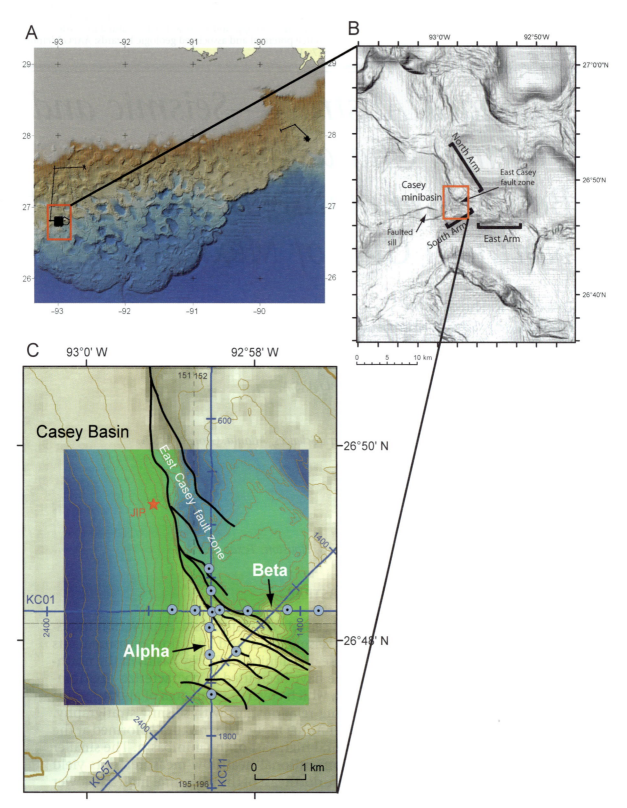

FIGURE 1. A) Shaded relief bathymetry of the Gulf of Mexico (taken from the National Oceanic and Atmospheric Administration coastal relief model); the box shows the location of panel B. B) Slope map of the area in the vicinity of the Casey minibasin; slope map created from the NOAA coastal relief model of the Gulf of Mexico; the box shows the location of panel C. C) Bathymetric map of the study area. Tan colors are shaded relief bathymetry from the NOAA coastal relief model for the Gulf of Mexico. The box shows bathymetry taken from 3-D multichannel seismic data. Locations of seismic lines (blue) and heat flow stations (blue circles), and the Chevron Joint Industry Project research well (red star) are shown, together with geologic features.

Comparisons were made of the results of several seismic experiments in the area of interest in MC798, including the prototype VLA test in 2003 and conventional multichannel and single-channel seismic data from previous years. In this study, these data sets were integrated to show the improved resolution of the near-surface sediments in the VLA data. Two interpretations of the geology are given; the evidence for the presence of subsurface gas hydrate is ambiguous.

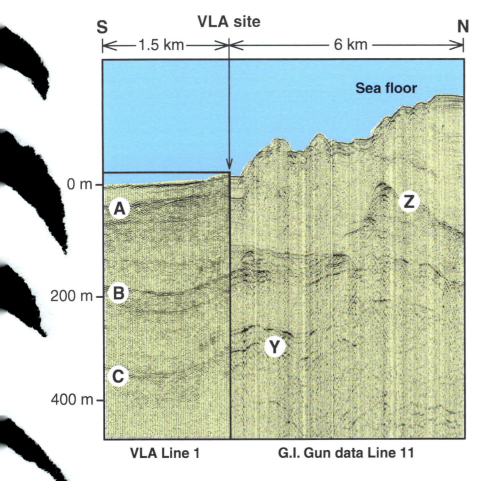

FIGURE 3. Direct comparison of the conventional 2-D seismic data (Gas-injection Gun data Line 11) and a selected, single common-receiver gather of the vertical line array data (VLA Line 1). A, B, and C represent the major reflectors and Y and Z show the wedge-shaped features.

Scholl, D. W., G. A. Barth, and J. R. Childs, 2009, Extended abstract—Why hydrate-linked velocity-amplitude anomaly structures are common in the Bering Sea Basin: A hypothesis, in T. Collett, A. Johnson, C. Knapp, and R. Boswell, eds., Natural gas hydrates—Energy resource potential and associated geologic hazards: AAPG Memoir 89, p. 29–31.

8

Extended Abstract—Why Hydrate-linked Velocity-amplitude Anomaly Structures are Common in the Bering Sea Basin: A Hypothe.'s

David W. Scholl, Ginger A. Barth, and Jonathan R. Childs
U.S. Geological Survey, Menlo Park, California, U.S.A.

ABSTRACT

The thick sedimentary sequence (2–12 km [6500–39,000 ft]) underlying the abyssal floors (3–4 km [9800–13,100 ft]) of the Bering Sea Basin is shallowly (<360 m [<1180 ft]) underlain by large (>2 km [>6500 ft] in diameter, ~200 m [65r ˇ ᵗhick) deposits of concentrated methane hydrate. Mound-shaped bodies of hyα. ⌐ displayed on seismic reflection records as VAMP (velocity-amplitude anomα. ˙ctures imaged as velocity pull-ups overlying push-downs. The VAMPs are nume˟. ˙undreds to thousands) and occur across an area of approximately 250,000 km ⁻25 mi²).

The abunσanc of VAMP structures is conjectured to be a consequence of high rates of basι. ˙ planktonic productivity, preservation of organic matter, biosiliceous sedimeι. ˙n, silica diagenesis, high heat flow, and deposition of a thick (700–1000 m [2. ˙300 ft]), upper section of perhaps latest Miocene but mostly glacial-age (early ɪ .ocene and Quaternary) turbidite beds and diatom ooze. Stacking of this upper Cenozoic sequence of water-rich beds heated underlying diatomaceous deposits of Miocene and older age and enhanced the generation of thermogenic methane and the diagenetic conversion of the opal A of porous diatom beds to the denser and contractionally fractured opal-cristobalite-tridymite phase of porcellaneous shale. Silica transformation expelled large volumes of interstitial and silica-bound water that, with methane, ascended through the shale via chimneys of fracture pathways to enter the porous (~60%) upper Cenozoic

DOI:10.1306/13201149M893346

section of diatom ooze and turbidite beds. Ascending methane entered the hydrate stability field at approximately 360 m (1180 ft), above which concentrated deposits of methane hydrate formed as either pore-filling accumulations or more massive lenses.

The deposition of high-velocity methane hydrate above a multitude of chimney structures transporting low-velocity, gas-charged fluids toward the seafloor is posited to account for the widespread recording of VAMP structures in the Bering Sea Basin.

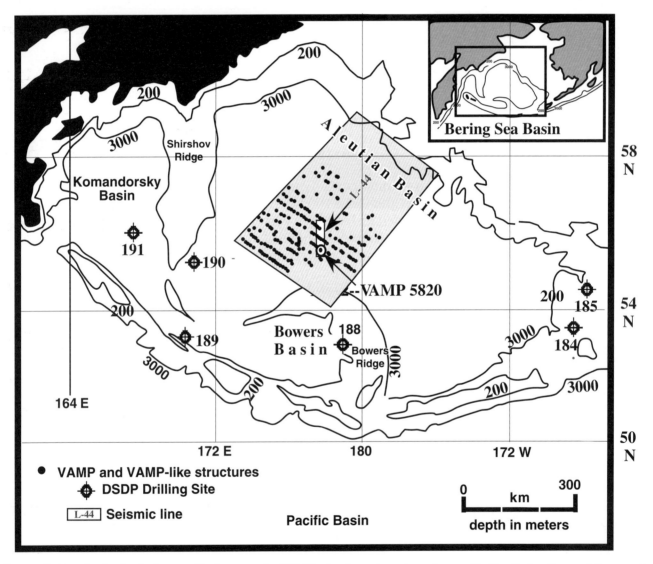

FIGURE 1. Hundreds of velocity-amplitude anomaly (VAMP) structures have been imaged in the Bering Sea and thousands are inferred to exist. The VAMP and VAMP-like structures recorded in the shaded box were identified by Rearic et al. (1988) along single-channel (SCS) lines collected in this sector of the U.S. Geological Survey (USGS) Bering Sea GLORIA survey of 1987. DSDP = Deep Sea Drilling Project; 1 km = 3281 ft.

VAMP 5820, Central Aleutian Basin

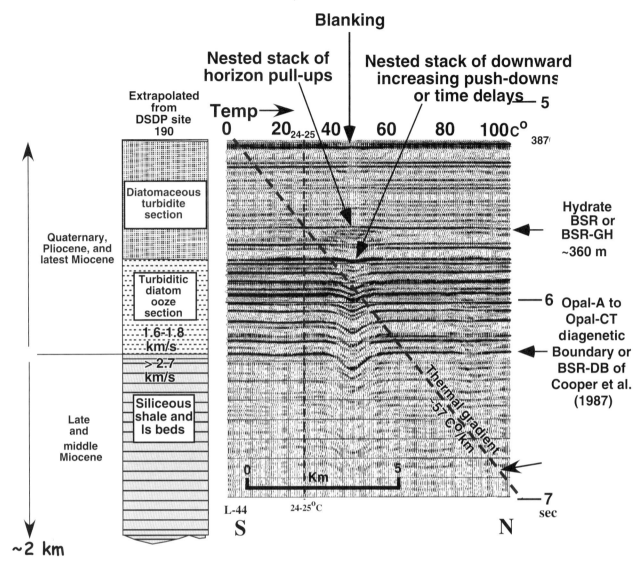

FIGURE 2. The multichannel (MCS) seismic section showing acoustic characteristics and stratigraphic, thermal, and diagenetic setting of velocity-amplitude anomaly (VAMP) structure 5820, which occurs in the central area of the Aleutian Basin along line L-44 (see location on Figure 1). Evidence is noticeable above the main hydrate body of acoustic blanking or muting of internal reflectivity. Blanking appears to be continuous to the seafloor, suggesting that venting of fluids may occur above VAMP structures. BSR-GH refers to a BSR (bottom-simulating reflector) linked to the upward transition from free pore-space gas to hydrated methane. At greater depth, the BSR-DB tracks the upward transition from diagenetically altered, cryptocrystalline siliceous shale below to overlying and unaltered, opaline-rich diatomaceous sediment. 1 km = 3281 ft.

REFERENCES CITED

Cooper, A. K., D. W. Scholl, and M. S. Marlow, 1987, Structural framework, sedimentary sequences, and hydrocarbon potential of the Aleutian and Bowers basins, Bering Sea, *in* D. W. Scholl, A. Grantz, and J. G. Vedder, eds., Geology and resource potential of the continental margin of western North America and adjacent ocean basins, Beaufort Sea to Baja California: Circum-Pacific Council for Energy and Mineral Resources, Houston, Texas, Earth Science Series 6, p. 473–502.

Rearic, D. M., S. R. Williams, P. R. Carlson, and R. K. Hall, 1988, Acoustic evidence for gas-charged sediment in the abyssal Aleutian Basin, Bering Sea, Alaska: U.S. Geological Survey Open-File Report 88-677, 40 p.

9

Barth, G. A., D. W. Scholl, and J. R. Childs, 2009, Extended abstract—Bering
Sea velocity-amplitude anomalies: Exploring the distribution of natural
gas and gas hydrate indicators, *in* T. Collett, A. Johnson, C. Knapp, and
R. Boswell, eds., Natural gas hydrates—Energy resource potential and
associated geologic hazards: AAPG Memoir 89, p. 33–36.

Extended Abstract—Bering Sea Velocity-amplitude Anomalies: Exploring the Distribution of Natural Gas and Gas-hydrate Indicators

Ginger A. Barth, David W. Scholl, and Jonathan R. Childs
U.S. Geological Survey, Menlo Park, California, U.S.A.

ABSTRACT

Velocity-amplitude anomalies (VAMPs), comprising coincident seismic travel-time anomalies and gas bright spots, are features widely identified in seismic reflection images from the deep-water Bering Sea basins. Interval traveltime anomalies are used to develop a method for the objective detection and quantification of these features. The approach selected uses relative traveltime variation in the sedimentary intervals above and below the gas-hydrate bottom-simulating reflector (BSR) as a diagnostic, measuring pull-up in the hydrate stability zone and push-down in the underlying gas zone, relative to a 400-common-depth-point (CDP) running-average interval reference. The method is used to explore the distribution of gas and hydrate indicators within a 120-km (74-mi) reflection profile segment in the central Aleutian Basin. This study segment includes 17 detected VAMPs, only 6 of which appear to contain significant quantities of stored hydrate. The total estimated volume of natural gas stored within the hydrate caps of these VAMPs is approximately 4 tcf (0.1 tcm). The largest three VAMP features contain greater than 85% of that total. Not all of the most visually obvious VAMP anomalies are important hydrate contributors. We suggest that VAMPs are fluid-expulsion features that have become involved in the transport of natural gas. As such, the VAMP systems should have been more active in the past. The VAMPs with significant hydrate present are most likely to be active today. The largest VAMP anomalies, including all of those associated with hydrate indicators, are located above prominent basement highs. This association suggests that fluid-migration patterns in these undeformed deep-water basins were originally established in response to sedimentation and compaction over basement topography, and that those ancient patterns have never been superceded. It also suggests that to make an informed estimate of gas-hydrate total volumes for the deep-water Bering Sea, the regional relationship between VAMP hydrate concentrations and basement topographic highs should be considered.

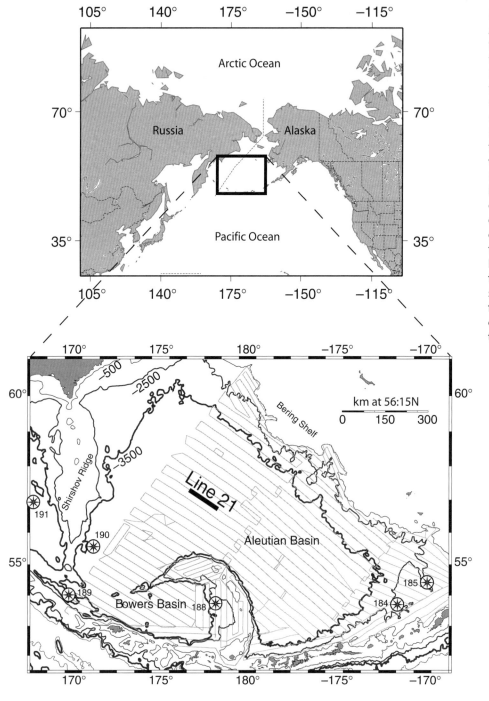

FIGURE 1. The deep-water Aleutian and Bowers basins of the Bering Sea lie north of the Aleutian Islands between Russia's Shirshov Ridge and the U.S. Bering Shelf. The U.S. Geological Survey single channel seismic reflection data (gray track lines) provide uniform basinwide coverage with an approximately 30-km (19-mi) line spacing. The black line marks the segment of the cruise *Farnella2-86-bs* line 21 evaluated in chapter 9, located on the CD-Rom of this publication. The Deep Sea Drilling Project (DSDP) site locations from Leg 19 are marked with stars (Sites 184–191). Results from Site 190 provide the most current and proximal ground truth information for our studies.

10

Paull, C. K., and W. Ussler III, 2009, Extended abstract—No evidence for enhanced methane flux from the Blake Ridge Depression, *in* T. Collett, A. Johnson, C. Knapp, and R. Boswell, eds., Natural gas hydrates—Energy resource potential and associated geologic hazards: AAPG Memoir 89, p. 37–38.

Extended Abstract—No Evidence for Enhanced Methane Flux from the Blake Ridge Depression

Charles K. Paull and William Ussler III
Monterey Bay Aquarium Research Institute, California, U.S.A.

ABSTRACT

Piston cores were collected from the floor, flanks, and background sediments associated with the Blake Ridge Depression off the East Coast of the United States to determine if this area is a gas-venting site. The hypothesis was that if the depression is associated with focused methane flux, authigenic carbonate mineralization, indicative of methane-related diagenesis and/or steeper pore water sulfate gradients, should occur within the feature. Sulfate gradients are sensitive to elevated pore-water methane concentrations and anaerobic methane oxidation. Compared with surrounding background sediments, neither steep sulfate gradients nor authigenic carbonates were observed within the cores collected from the interior of the Blake Ridge Depression. The hypothesis that the Black Ridge Depression is or has been a site of enhanced methane venting is not supported by these geochemical observations.

REFERENCE CITED

Dillon, W. P., J. W. Nealon, M. H. Taylor, M. W. Lee, and R. M. Drury, 2001, Seafloor collapse and methane venting associated with gas hydrate on the Blake Ridge—Causes and implications to sea floor stability and methane release, *in* C. K. Paull and W. P. Dillon, eds., Natural gas hydrates, occurrence, distribution and detection: American Geophysical Union Geophysical Monograph 124, p. 211–234.

Nealon, J. W., W. P. Dillon, W. Danforth, T. F. O'Brien, 2002, Deep-towed chirp profiles on the Blake Ridge collapse structure collected aboard R/V *Cape Hatteras* in 1992 and 1995 (CD-ROM): U.S. Geological Survey Open-File Report 01-123, CD-ROM.

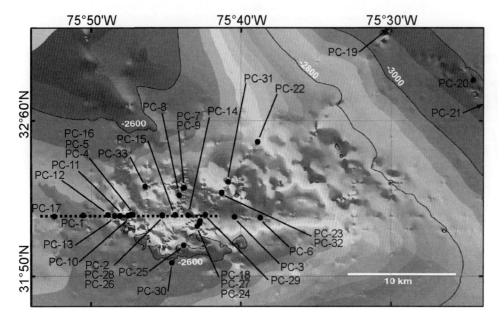

FIGURE 1. Bathymetric map of the Blake Ridge crest from Dillon et al. (2001) showing the irregular topography referred to as the Blake Ridge Depression (BRD). The locations of piston cores are indicated. The dotted line shows the location of the seismic profile shown in Figure 2. 1 km = 0.6 mi.

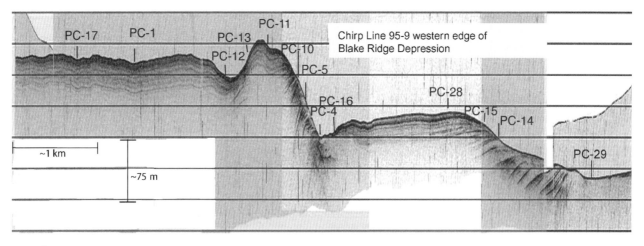

FIGURE 2. West–east-oriented deep-towed chirp seismic reflection profile running across the Blake Ridge Depression showing an outcropping reflector on the scars on the side of the depression. Data collected by the U.S. Geological Survey (Nealon et al., 2002). The locations of piston cores and seismic profile are shown in Figure 1.

Janik, A., D. Goldberg, G. Guerin, and T. Collett, 2009, Extended abstract—
Estimation of gas-hydrate saturation and heterogeneity on Cascadia
Margin from Ocean Drilling Project Leg 204 logging-while-drilling
measurements, in T. Collett, A. Johnson, C. Knapp, and R. Boswell, eds.,
Natural gas hydrates—Energy resource potential and associated geologic
hazards: AAPG Memoir 89, p. 39–40.

11

Extended Abstract—Estimation of Gas-hydrate Saturation and Heterogeneity on Cascadia Margin from Ocean Drilling Project Leg 204 Logging-while-drilling Measurements

Aleksandra Janik[1], David Goldberg, and Gilles Guerin
Columbia University, Lamont Doherty Earth Observatory, Palisades, New York, U.S.A.

Timothy Collett
U.S. Geological Survey, Denver, Colorado, U.S.A.

ABSTRACT

In this study, we present an improved method of estimation of pore space saturated with gas hydrate from inversion of resistivity and porosity data. The method is developed through analysis of logging data collected during logging-while-drilling (LWD) measurements on Hydrate Ridge offshore Oregon, whereby the three-dimensional distribution of hydrate is explicitly considered. The LWD tools rotate and provide 56 resistivity and 16 density azimuthal measurements of the sediment properties around the inner circumference of the borehole collected with high vertical resolution (3 cm [1.1 in.]). Additionally, LWD data are acquired only minutes after the formation is drilled, limiting the extent of hydrate dissociation in the measured in-situ properties. The resulting borehole wall images of gas-hydrate saturation reveal the detail of the geometry and shape of void spaces in which gas hydrate occurs. Several visible fractures are filled with gas hydrate, which also occur with patchy distribution, in lenticular shapes, as asymmetrical fracture fill, and in truncated thin layers. The LWD data indicate

[1]*Present address*: Exxon Mobil Exploration Company, Houston, Texas.

DOI:10.1306/13201152M893349

Resistivity images and low temperature anomalies

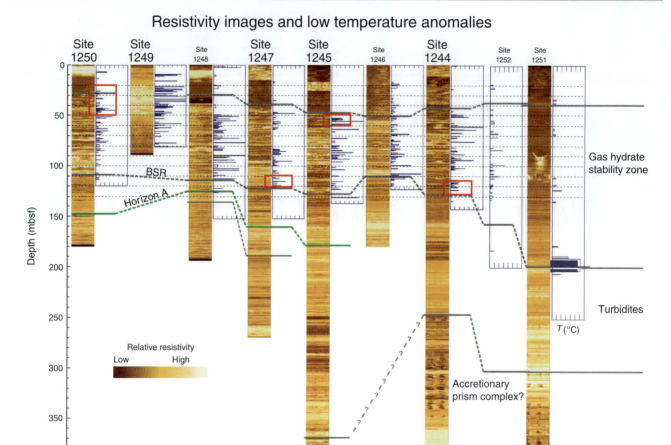

FIGURE 1. Relative borehole resistivity as imaged by logging while drilling (LWD) (golden or brown columns) compared to the presence of low-temperature anomalies in recovered cores as imaged by infrared (IR) camera scans (blue spikes). Depth scale is in meters below seafloor (mbsf). Light shades represent relatively higher resistivity. The IR anomalies represent the difference between local and background temperature. Red rectangles indicate areas where the LWD deep-button resistivity confirms the presence of gas hydrate, whereas estimates based on ring resistivity yield 0% saturation (modified from Trehu et al., 2003). 1 m = 3.28 ft.

considerable azimuthal heterogeneity at the Hydrate Ridge sites, with typical variations of more than four times the azimuthal average of saturation at a given depth and on a centimeter (inch) scale around the borehole. Such heterogeneous and patchy hydrate distribution is characteristic for all occurrences of hydrate within the gas-hydrate stability zone at all of the investigated Hydrate Ridge sites.

REFERENCE CITED

Trehu, A. M., G. Bohrmann, F. R. Rack, M. E. Torres, et al., 2003, Proceedings of the Ocean Drilling Program: Initial reports: College Station, Texas, Ocean Drilling Program, v. 204, doi:10.2973/odp.proc.ir.204.2003.

12

Fujii, T., M. Nakamizu, Y. Tsuji, T. Namikawa, T. Okui, M. Kawasaki, K. Ochiai, M. Nishimura, and O. Takano, 2009, Extended abstract—Methane-hydrate occurrence and saturation confirmed from core samples, eastern Nankai Trough, Japan, *in* T. Collett, A. Johnson, C. Knapp, and R. Boswell, eds., Natural gas hydrates—Energy resource potential and associated geologic hazards: AAPG Memoir 89, p. 41–43.

Extended Abstract—Methane-hydrate Occurrence and Saturation Confirmed from Core Samples, Eastern Nankai Trough, Japan

Tetsuya Fujii, Masaru Nakamizu, and Yoshihiro Tsuji
Technology and Research Center, Japan Oil, Gas and Metals National Corporation (JOGMEC), Japan

Takatoshi Namikawa
Japan Petroleum Exploration Co., Ltd., Japan

Toshiharu Okui
Gas Resources Department, Tokyo Gas Co., Ltd., Japan

Masayuki Kawasaki
Japan Drilling Co., Ltd. (JDC), Japan

Koji Ochiai
INPEX Corporation, Japan

Mizue Nishimura and Osamu Takano
Research Center, Japan Petroleum Exploration Co., Ltd., Japan

ABSTRACT

The Ministry of Economy, Trade and Industry, Japan, drilled the Tokai-oki to Kumano-nada exploratory test wells to obtain data for understanding the occurrence of methane hydrate and estimating the volume of gas stored as methane hydrates in the Nankai Trough, offshore central Japan. In this project, we conducted logging-while-drilling at 16 sites, coring at 4 sites, wireline logging at 2 sites, and long-term monitoring of formation temperature at a single site.

Massive or layered methane hydrates within muddy layers were recovered at sites 1 and 2 by drilling with a conventional wireline-core system. The methane-hydrate-bearing sediments in these sites are a combination of clay and silt, which

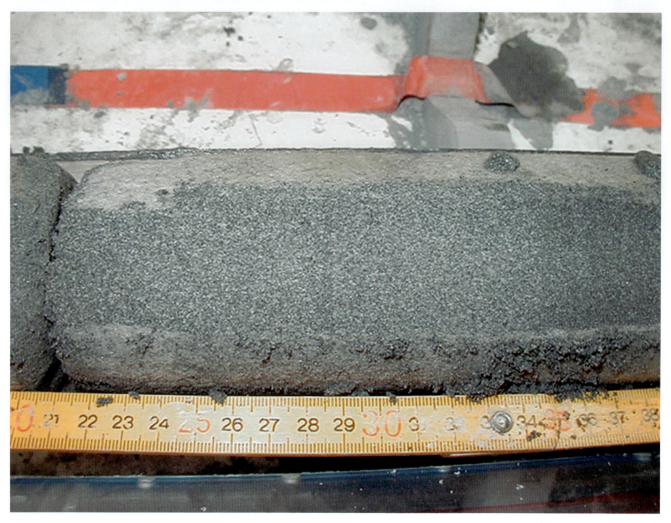

FIGURE 1. Pore space hydrate in sand layers recovered by pressure-temperature core sampler (PTCS) coring at site 13. The sample was recovered from 164.3 meters below sea floor (mbsf) by PTCS coring (well no. 25). White small particles are gas hydrates.

is not commonly considered a favorable host sediment for hydrate formation; however, a significant decrease in core temperature was recorded within intervals of layered hydrates.

Pore-space-type hydrates were identified in sand layers from sites 4 and 13 within an 82-m (269-ft) core recovered using a pressure-temperature core sampler (PTCS). Sediments within this core are mainly very fine- to fine-grained turbidite sand layers of several centimeters (inches) to 1.5 m (5 ft) in thickness (average of 20–40 cm [8–16 in.]). Core-temperature measurements and the relationship between well-log resistivity and grain-size distribution indicate that methane hydrate is concentrated within layers of coarse-grained sand. We identify five sedimentary facies on the basis of a lithological column created from core and well-log data at sites 4 and 13. Facies analysis indicates that the depositional environment in hydrate-bearing zones at sites 4 and 13 consisted of distributary channels to distal lobes within a submarine-fan system. Shipboard hydrate dissociation tests on PTCS cores reveal that average hydrate saturation in the cored sand layers ranged from 55 to 68%, with the average sediment porosities ranging from 39 to 41%, based on an analysis of both well log and core measurements.

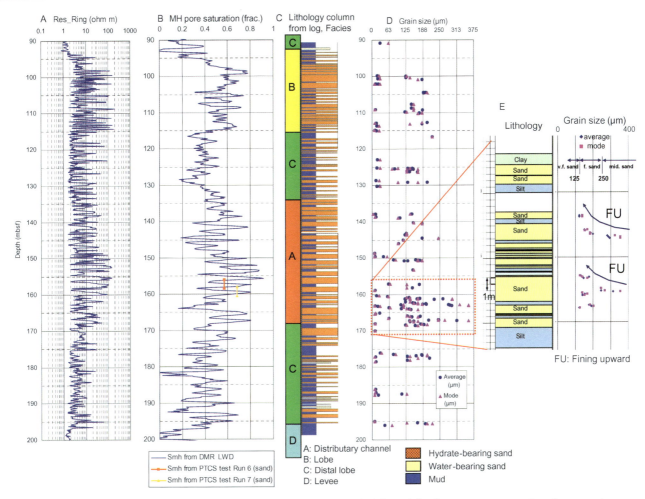

FIGURE 2. Depth plot of A) logging-while-drilling (LWD) ring resistivity, B) hydrate pore saturation from nuclear magnetic resonance (NMR) logs and the pressure-temperature core sampler (PTCS) dissociation test, C) lithological columns created from well logs and facies interpretation, D) the results of grain size analysis, and E) identified fining-upward characteristics for site 13. MH = methane hydrate; DMR = density magnetic resonance method; mbsf = meters below sea floor; Smh = methane hydrate saturation; 10 m = 32.8 ft.

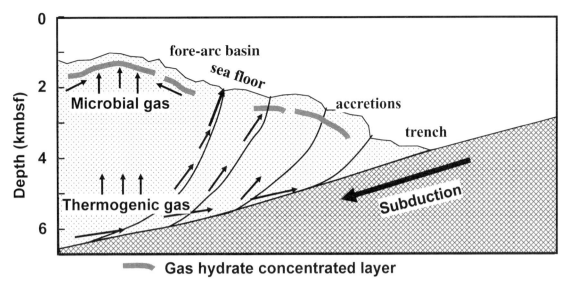

FIGURE 2. Schematic diagram showing possible gas migration and gas hydrate formation in Nankai Trough, southeast Japan. Gas hydrates in the Nankai Trough contain mainly microbial methane, and the migration distance is rather short in the Nankai Trough. kmbsf = kilometers below sea floor; 1 kmbsf = 3281 ft below sea floor.

controls on the occurrence of conventional oil and gas accumulations. We suggest the distribution of porous and coarser grained host sandy sediments is one of the most important factors controlling the occurrence of gas hydrates, as well as pressure-temperature conditions. Based on the isotopic and molecular compositions, marine gas hydrates in the Nankai Trough contain mainly microbial methane, whereas terrestrial gas hydrates in the Mackenzie delta contain thermogenic gas. Although selective migration of methane to sandy layers is observed, the migration distance is rather short in the Nankai Trough, whereas the long migration of thermogenic gas generated in much deeper and/or more mature sediments through faults is inferred in the Mackenzie delta.

REFERENCE CITED

Dixon, J. D., G. R. Morrell, J. R. Dietrich, G. C. Taylor, R. M. Procter, R. F. Conn, S. M. Dallaire, and J. A. Christie, 1994, Petroleum resources of the Mackenzie delta and Beaufort Sea: Geological Survey of Canada Bulletin, v. 474, 86 p.

Fukuhara, M., V. Tertychnyi, K. Fujii, V. Shako, V. Pimenov, Y. Popov, D. Murray, and T. Fujii, 2009, Extended abstract—Temperature monitoring results for methane-hydrate sediments in the Nankai Trough, *in* T. Collett, A. Johnson, C. Knapp, and R. Boswell, eds., Natural gas hydrates—Energy resource potential and associated geologic hazards: AAPG Memoir 89, p. 49–51.

14

Extended Abstract—Temperature Monitoring Results for Methane-hydrate Sediments in the Nankai Trough, Japan

Masafumi Fukuhara and Vladimir Tertychnyi
Schlumberger Moscow Research Center, Moscow, Russia

Kasumi Fujii
Schlumberger K.K., Kanagawa, Japan

Valery Shako, Viacheslav Pimenov, and Yuri Popov
Russian State Geological Prospecting University, Moscow, Russia

Doug Murray
Schlumberger China S.A., Beijing, China

Tetsuya Fujii
Japan Oil, Gas and Metals National Corporation, Chiba, Japan

ABSTRACT

The Japan Oil, Gas and Metals National Corporation (JOGMEC) conducted the Kisoshisui (Tokaioki-Kumanonada) drilling campaign in the Nankai Trough area during 2003–2004 as part of the activities of the MH21 Research Consortium, which is supported by the Ministry of Economy, Trade and Industry (METI). Through this campaign, various subsea hydrate studies were conducted using various data including logging while drilling, wireline, coring, and temperature measurement. Using recent technology, a precise in-situ temperature measurement system was successfully developed and deployed at the beginning of the drilling campaign. The goals of the study were to determine whether the measurement system and setup can reliably measure the equilibrium in-situ formation temperature, to determine if the measurements adequately delineate the geothermal gradient and to identify any disturbances in the thermal regime of the

DOI:10.1306/13201155M893352

hydrate-bearing sediment and the possible causes of the disturbances. The temperature of the hydrate-bearing sediment was monitored in a deep-water well for 1.5 months. Because the measurements were collected over a long period of time, the equilibrated formation temperatures could be readily determined after observing the thermal relaxation from drilling and sensor deployment effects. The temperature profile and the temperature gradient through the hydrate-bearing zone were estimated in a quasi-steady-state condition. A correlation between the tidal cycle and sea-floor temperature change is also discussed.

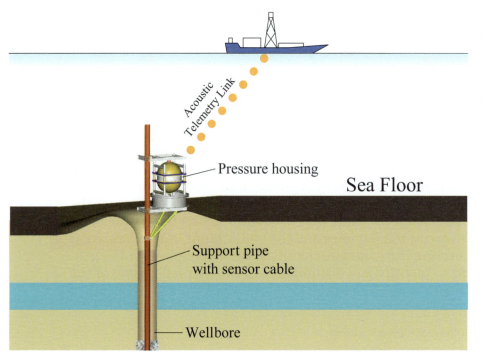

FIGURE 1. A precise in-situ temperature measurement system employing fiber optic technology was successfully deployed at the beginning of the drilling campaign.

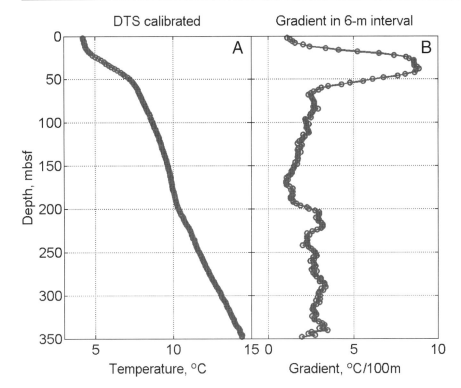

FIGURE 2. A) Subsurface depth plot of the averaged temperature profile as measured by the fiber optic system and B) the estimated temperature gradient from the measured temperature profile. DTS = distributed temperature measurement system; mbsf = meters below sea floor; 1 m = 3.28 ft.

Bünz, S., J. Mienert, and K. Andreassen, 2009, Extended abstract—
Multicomponent seismic studies of the gas-hydrate system at the
Storegga Slide, in T. Collett, A. Johnson, C. Knapp, and R. Boswell, eds.,
Natural gas hydrates—Energy resource potential and associated
geologic hazards: AAPG Memoir 89, p. 57–59.

16

Extended Abstract—Multicomponent Seismic Studies of the Gas-hydrate System at the Storegga Slide

Stefan Bünz, Jürgen Mienert, and Karin Andreassen
Department of Geology, University of Tromsø, Tromsø, Norway

ABSTRACT

A multicomponent seismic technology is able to broaden our knowledge of the gas-hydrate reservoir. In the marine environment, shear waves (S waves) can be generated by conversion from a downward-propagating compressional wave (P wave) upon reflection at a sedimentary interface. The upward-propagating S wave can be recorded at the ocean floor using horizontal geophones. S waves can be useful in addition to P-wave data because the S-wave velocity is slower than P-wave velocity, and S waves are less affected by pore fill of porous rocks. This clearly gives a distinct improvement because (1) seismic resolution using S waves increases, (2) targets of gas or of poor P-wave reflectivity are imaged well, (3) pore fluids and lithology can be discriminated, and (4) the enhanced ability exists to estimate gas-hydrate concentrations. On the mid-Norwegian margin, multicomponent seismic data have enabled us to choose a proper rock-physics model for the hydrate-bearing sediments. We are able to constrain seismic velocities from ocean-bottom seismic data. This allows us to obtain more accurate estimations of gas-hydrate and free-gas concentrations and to assess the occurrence of overpressures within the gas-bearing sediments underneath the hydrates. Improved acoustic images look through the zone underneath the hydrate-bearing sediments, which is obscured on the P-wave data because of the occurrence of gas.

DOI:10.1306/13201157M893354

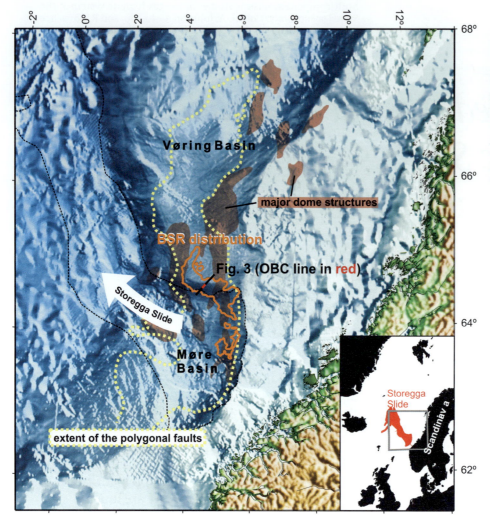

FIGURE 1. Distribution of the bottom-simulating reflectors (BSRs) and pockmarks or pipes on the mid-Norwegian margin from Bünz et al. (2003). The extent of polygonal faults from Berndt et al. (2003). The major dome structures are from Vagnes et al. (1998). The location of the ocean-bottom cable (OBC) line lies within the BSR area on the northern flank of the Storegga Slide.

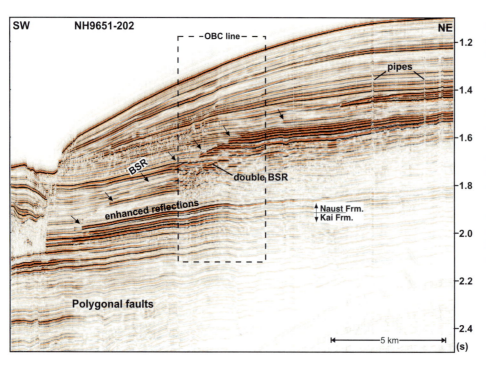

FIGURE 2. Time-migrated seismic section from the northern flank of the Storegga Slide. The bottom-simulating reflectors (BSRs) on the mid-Norwegian margin are mainly identified as the termination of enhanced reflections. A polygonal fault system occurs in the sediments of the Kai Formation. Pipes can be identified on the upper part of the slope between the Vøring and the Møre basins. Underneath them is a seismically transparent zone. The dashed box marks the location of the ocean-bottom cable (OBC) line. 1 km = 0.6 mi.

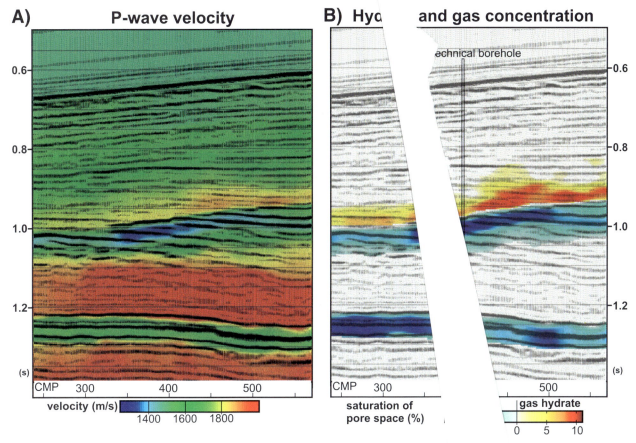

FIGURE 3. A) Wiggle-trace plot of the vertical component overlain with color-code   ional wave (P-wave) velocities. The bottom-simulating reflector (BSR) can be clearly identified by a veloci   . B) Wiggle-trace plot of the vertical component with hydrate and gas saturations of pore space calculated fro   te-as-part-of-frame and the homogeneous-gas-distribution rock-physics model of Helgerud et al. (1999). Hy   as saturations vary considerably along the line. The rectangle marks the location of the geotechnical bo   ' = common mid point.

REFERENCE CITED

Berndt, C., S. Bünz, and J. Mienert, 2003, Polygonal fault systems on the mid-Norwegian margin: A long term source for fluid flow, *in* P. Van Rensbergen, R. Hillis, A. Maltman, and C. Morley, eds., Subsurface sediment mobilization: Geological Society (London) Special Publication 216, p. 283–290.

Bünz, S., J. Mienert, and C. Berndt, 2003, Geological controls on the Storegga gas-hydrate system of the mid-Norwegian continental margin: Earth and Planetary Science L   307.

Helgerud, M. B., J. Dvorkir   1999, Elastic-wave vel   gas hydrates: Effective   Research Letters, v. 26,

Vagnes, E., R. H. Gabrielse   Cretaceous–Cenozoic i   mation at the Norwegi   magnitude and regiona   ics, v. 12, no. 1–4, p. 29

9, no. 3–4, p. 291–

.akai, and T. Collett,   ine sediments with   leling: Geophysical   21–2024.

.remo, 1998, Late   ntractional defor-   tal shelf, timing,   is: Tectonophys-

17

Fohrmann, M., A. R. Gorman, and I. A. Pecher, 2009, Extended abstract—Seismic characterization of the Fiordland gas-hydrate province, New Zealand, *in* T. Collett, A. Johnson, C. Knapp, and R. Boswell, eds., Natural gas hydrates—Energy resource potential and associated geologic hazards: AAPG Memoir 89, p. 61–63.

Extended Abstract—Seismic Characterization of the Fiordland Gas-hydrate Province, New Zealand

Miko Fohrmann[1] and Andrew R. Gorman

Department of Geology, University of Otago, Dunedin, New Zealand

Ingo A. Pecher

Institute of Geological and Nuclear Sciences Limited (GNS Science), Lower Hutt, New Zealand

ABSTRACT

Occurrences of bottom-simulating reflectors (BSRs), related to the presence of gas hydrates, have previously been observed across a widespread zone on the active continental margin associated with the incipient Puysegur subduction zone, southeast of the South Island of New Zealand. However, unlike New Zealand's other large gas-hydrate province located on the active Hikurangi margin, east of the North Island, the Fiordland BSRs have not been described in terms of their seismic characteristics or distribution. Five seismic reflection data sets are analyzed here to identify a region of BSRs covering approximately 2200 km^2 (849 mi^2). The BSRs identified in this region exhibit classic characteristics indicative of a reflector at the base of the gas-hydrate stability zone: (1) they predominantly have a negative polarity, implying a decrease in acoustic impedance; (2) they crosscut strata; and (3) they have a variable amplitude-with-offset (AVO) response, indicating the presence of free gas below the reflector. Localized regions of acoustic blanking may be observed at some points above strong BSRs. The lack of information on the sedimentary characteristics of the Fiordland margin limits our ability to quantify the gas-hydrate deposits in this province. However, a significant proportion (16%) of the mapped region contains structural and stratigraphic features that can focus the upward flow of fluids and may therefore correspond to regions of increased hydrate concentrations.

[1]*Present address*: Institute of Geological and Nuclear Sciences Limited (GNS Science), Lower Hutt, New Zealand.

DOI:10.1306/13201158M893355

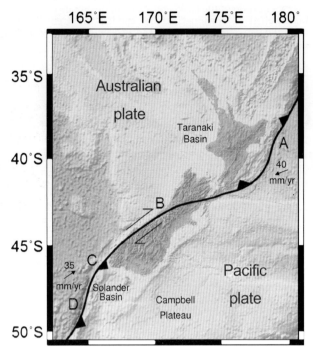

FIGURE 1. Tectonic setting of New Zealand, showing the boundary of the Australian and Pacific plates. The westward-directed subduction of the Pacific plate beneath the Australian plate at the Hikurangi margin A) changes into strike-slip motion along the Alpine fault B) and further into eastward-directed subduction starting approximately at Resolution Ridge C) on the Puysegur Trench D). Arrows indicate the Pacific-Australian relative-plate-motion vector and rate (De Mets et al., 1994). The Campbell Plateau is located south of New Zealand. The research area is located on the continental slope southwest of the South Island, west of the Solander Basin. The Taranaki Basin, located on the passive margin, east of the North Island is New Zealand's only producing gas basin, whereas the petroleum potential of the Campbell Plateau has not been developed so far. 1 mm = 0.4 in.

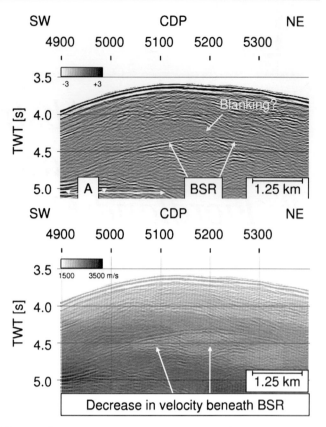

FIGURE 2. (top) Section of the *Geco Resolution* data set across Dusky Ridge, showing a strong botttom-simulating reflector (BSR) below the structural high. The subducting Australian Plate is marked with an 'A'. (bottom) Interval velocity distribution plotted on top of the seismic section shows a decrease of interval velocity below the BSR. The scale represents interval velocities between 1500 m/s (4921 ft/s) (white) and 3500 m/s (11,483 ft/s) (black). CDP = common depth point. TWT = two-way-time (s). 1 km = 0.6 mi.

REFERENCE CITED

De Mets, C., R. G. Gordon, D. F. Argus, and S. Stein, 1994, Effect of recent revisions to the geomagnetic reversal time-scale on estimates of current plate motions: Geophysical Research Letters, v. 21, p. 2191–2194.

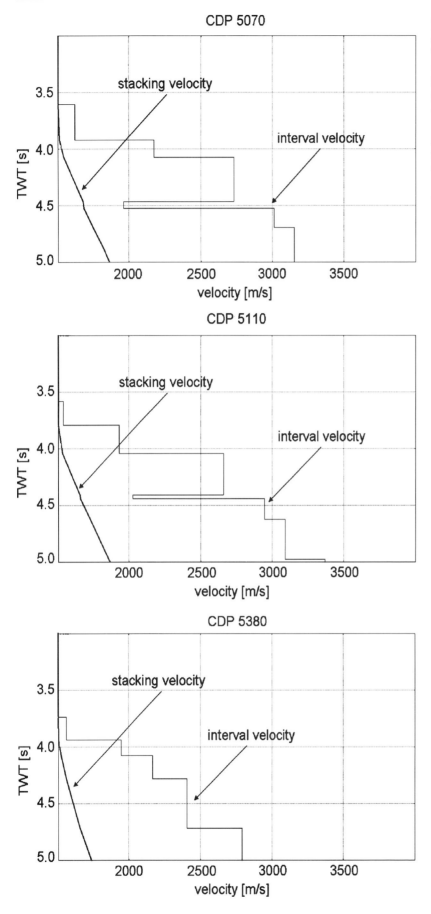

FIGURE 3. Three representative velocity distributions across Dusky Ridge. (top) Velocity profile of common depth point (CDP) 5070, showing an overall increase in stacking velocity with depth interrupted by a sharp drop in interval velocity below the bottom-simulating reflector (BSR). (middle) CDP 5110 exhibits a similar velocity distribution. (bottom) CDP 5380 reveals interval velocities that are lower than those to the west. Further, in the absence of a BSR, no drop in velocity is observed. TWT = two-way-time (s).

18

Henrys, S. A., D. J. Woodward, and I. A. Pecher, 2009, Extended abstract—
Variation of bottom-simulating-reflection strength in a high-flux methane
province, Hikurangi margin, New Zealand, *in* T. Collett, A. Johnson, C.
Knapp, and R. Boswell, eds., Natural gas hydrates—Energy resource
potential and associated geologic hazards: AAPG Memoir 89, p. 65–68.

Extended Abstract—Variation of Bottom-simulating-reflection Strength in a High-flux Methane Province, Hikurangi Margin, New Zealand

Stuart A. Henrys, Derek J. Woodward, and Ingo A. Pecher[1]

Institute of Geological and Nuclear Sciences (GNS Science), Lower Hutt, New Zealand

ABSTRACT

The Hikurangi margin, New Zealand, is well imaged offshore from seismic reflection data as a 100–150-km (62–93-mi)-wide accretionary prism that encroaches onshore in both northern and southern parts (Figure 1). It comprises a backstop of Mesozoic greywacke rocks and pre-existing passive margin sediments that range in age from middle Cretaceous to Paleogene and have been thrust faulted and back tilted during the past 25 m.y. Neogene slope basin sediments, which accumulated behind thrust ridges, have been progressively incorporated into the subduction complex. Trench-fill turbidites, filling these basins, are being accreted to the front of the Hikurangi margin at a rate of 12 ± 3 mm (0.5 ± 0.1 in.)/yr. Accretionary wedge sediments are estimated to contain approximately 25–38% pore water and expel fluid at greater than 5 m^3 (176 ft^3) annually for each meter along strike. Fluid-pressure trends from seven oil exploration wells approach the lithostatic trend, and manifestations of fluid are ubiquitous along the margin. Seeps and springs are found onshore and offshore where vents are marked by distinctive chemosynthetic faunas, carbonate chimneys, bubbles in the water column, and high methane concentrations.

Bottom-simulating reflections (BSRs) represent the base of a gas-hydrate zone underlain by widespread free gas. On the southern Hikurangi margin, multichannel seismic data reveal that the gas-hydrate province extends from about 600-m (1968-ft)

[1]*Present address*: Institute of Petroleum Engineering, Heriot-Watt University, Edinburgh, United Kingdom.

water depth to the Hikurangi trench (Figure 1a) and covers an area of about 50,000 km^2 (19,305 mi^2). We analyzed BSR strength (amplitude ratio of the BSR to sea floor and BSR coefficients) in a grid of seismic data across this area (Figures 1b and 2). Simplified rock-physics models were used to estimate the reflection coefficient of BSRs with a gas concentration above which compressional wave velocity is mostly insensitive to gas saturation. This reflection coefficient was found to be −0.20, resulting from at least 8–10% gas saturation. Four percent of the gas-hydrate stability zone (GHSZ) on the southern Hikurangi margin is underlain by strong BSRs with reflection coefficients that are −0.20 or stronger. Mapped variations in BSR and sea-floor reflection amplitude ratios and reflection coefficients point to a close association of strong BSRs (i.e., BSR coefficient <−0.2) with bathymetric highs and crosscutting strata, both features that enhance fluid flow across the base of the gas-hydrate stability. In contrast, regions of flat-lying sediments with no visible fault structures and crosscutting strata appear to have low BSR strength. The bathymetry of the Hikurangi margin accretionary prism is dominated by anticlinal ridges cored by thrust faults and steeply dipping strata, many of which are likely to act as conduits that promote fluid flow. Regionally, we propose that the gas for hydrate formation is migrating upward in solution along these permeable layers and pathways. Additional concentrations of gas beneath anticlines may occur on a local scale if free gas may be trapped at the base of the GHSZ where a permeability barrier is caused by hydrate clogging and migrates laterally toward topographic highs beneath anticlines. In this case, there must be sufficient concentration of gas hydrate to effectively decrease permeability. However, close to the interpreted faults, we speculate that locally, on a scale of several meters, increased permeability along faults may provide warmed channels, allowing the escape of methane through the GHSZ and to the sea floor.

Isotope, geochemical, and geophysical data from previous studies onshore point to a thermogenic origin for methane and suggest that New Zealand east coast fluids are derived from accreted, organic-rich, sedimentary sources overlying the subducting slab and that these sources must have an age of about 70 Ma. We therefore speculate that BSR formation on the Hikurangi margin is supported by long-term recycling of fluids along faults that penetrate through the oldest sediments in the forearc and sole at the plate interface, as mapped in crustal seismic sections elsewhere on the Hikurangi margin. Squeezed subducting sediments at the plate interface may provide a rich source of water driving fluid recycling.

Based on the results of this study, we have narrowed down potential study areas for more detailed gas hydrate surveys. Additional criteria for the selection of sites included a correlation of BSR strength with fluid flow features and the presence of known vent sites.

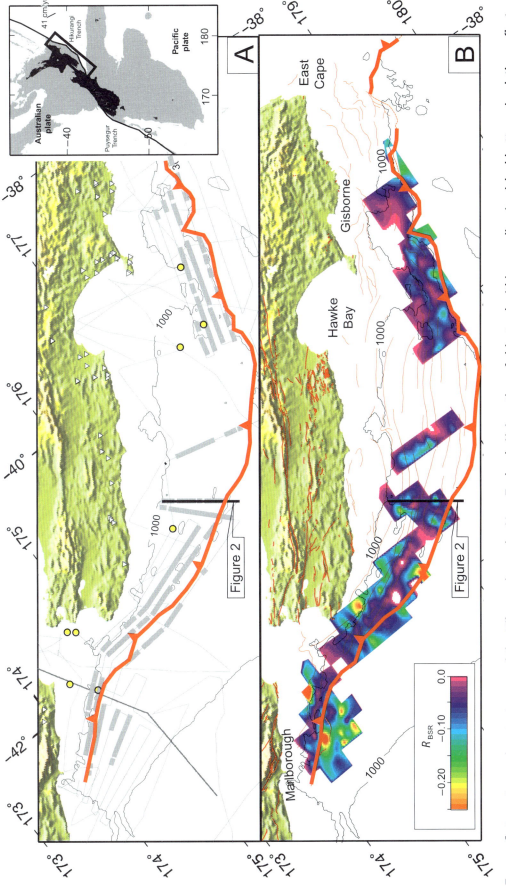

FIGURE 1. Location and tectonic map of the Hikurangi margin, New Zealand. A) Location of ship tracks (thin gray lines), picked bottom-simulating reflectors (BSRs) along track lines (filled gray boxes), known offshore seeps, and chemosynthetic fauna (filled circles), and sites onshore where gas and fluid samples have been analyzed (inverted open triangles). The thick tooth line marks the location of the frontal thrust along the Hikurangi margin. B) Reflection coefficient of the BSR. The BSRs are prevalent along most of the East Coast margin in water depths ranging from 600 to 3000 m (1968 to 9842 ft) and covers an area of about 50,000 km² (19,305 mi²). Thin black lines are active faults. Contours are water depth in meters.

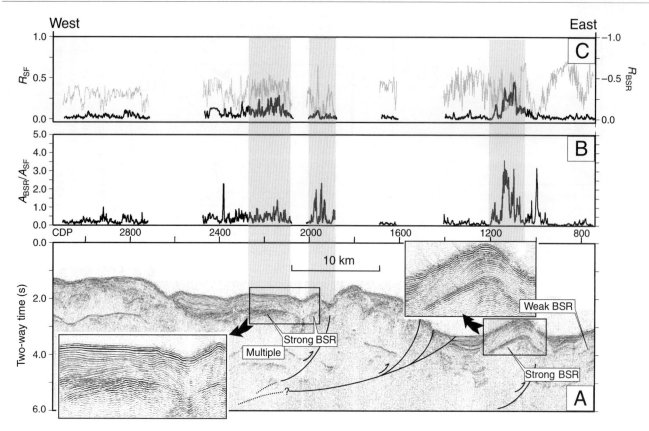

FIGURE 2. A) Migrated seismic data for line GeodyNZ36a. Two areas of identified strong bottom-simulating reflectors (BSRs) are enlarged. B) Amplitude ratio of A_{BSR}/A_{SF} where A_{BSR} is the amplitude of BSR and A_{SF} is the amplitude of the sea floor. C) Reflection coefficients of the sea floor (R_{SF}, gray) and BSR (R_{BSR}, black). Both A_{BSR}/A_{SF} and the reflection coefficients of the BSR show correlation to the areas of strong BSRs (represented by vertically shaded strip) visible in the migrated seismic image. The mapped distribution of R_{BSR} points to a close association of strong BSRs (i.e., $R_{BSR} < -0.2$) with bathymetric highs and crosscutting strata, both features that enhance fluid flow across the base of the gas-hydrate stability zone. In contrast, regions of flat-lying sediments with no visible fault structures and crosscutting strata appear to have low A_{BSR}/A_{SF} ratios and R_{BSR} approximately greater than -0.1 (i.e., a weaker reflection coefficient).

Schnurle, P., and C.-S. Liu, 2009, Extended abstract—Structural controls on
the formation of bottom-simulating reflectors offshore southwestern
Taiwan from a dense seismic reflection survey, *in* T. Collett, A. Johnson,
C. Knapp, and R. Boswell, eds., Natural gas hydrates—Energy resource
potential and associated geologic hazards: AAPG Memoir 89, p. 69–72.

19

Extended Abstract—Structural Controls on the Formation of Bottom-simulating Reflectors Offshore Southwestern Taiwan from a Dense Seismic Reflection Survey

Philippe Schnurle and Char-Shine Liu

Institute of Oceanography, National Taiwan University, Taipei, Taiwan

ABSTRACT

A dense seismic reflection survey with a 400-m (1312-ft) line spacing has
been conducted in a 14- by 16-km (8- by 10-mi) area offshore southwestern
Taiwan where bottom-simulating reflectors (BSRs) are highly concentrated and geochemical signals for the presence of gas hydrate are strong. A
complex interplay between north–south-trending thrust faults and northeast–
southwest oblique ramps exists in this region, which impacts the distribution and
seismic characteristics of the BSRs. A clear BSR is observed almost continuously in
the southeastern half of the survey box, whereas BSRs either appear as small
patches or are absent in the northwestern half. The reflection coefficients at the
BSRs presents abrupt lateral variations, with an average value of -0.094 or 32% of
the reflection coefficient at the sea floor. A pattern of seismic blanking beneath
the ridge crests overlying a high-amplitude BSR and bright reflectors below the
BSRs in the slope basins reveal the significance of tectonics and sedimentation
of the formation of BSRs. Local heat-flow values estimated from BSR subbottom
depths show different ranges across the northeast–southwest-trending Yung-An
lineament, with higher heat-flow values ranging from 55 to 70 mW/m^2 in the
southeastern half of the survey area, whereas the heat-flow values range from 45 to
55 mW/m^2 to the northwest of the Yung-An lineament. In the southeastern part
of the survey area, an elongated diapiric feature deforms the sedimentary layers,
BSRs occur shallower, and high heat-flow values are inferred. Furthermore, local
dips of BSRs and sedimentary strata that crosscut the BSRs at intersections of any
two seismic profiles have been computed. A dominant N308° stratigraphic updip

direction characterizes the anticlinal ridges, and strata dips fan out toward N285° south of the ridges and toward N330° north of the ridges. The concentric patterns of steep stratigraphic dips constitute favorable migration paths for the upward transport of fluids with significant dissolved-gas content across the BSR, which correlates well with the high gas volumes sampled in this area.

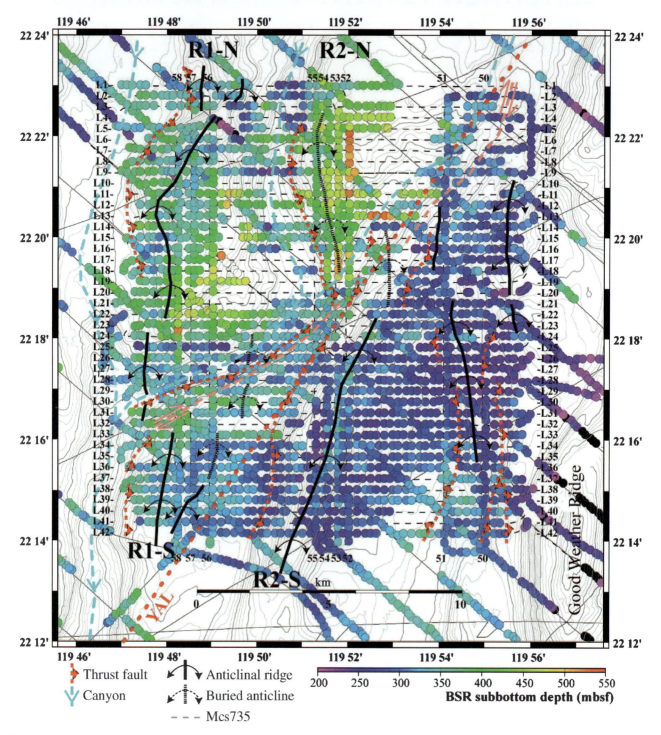

FIGURE 1. Structural map, bottom-simulating reflector (BSR) distribution, and subbottom depth in the dense survey area. The dominant structures are north–south-trending thrust-bounded folds separated by northeast–southwest oblique ramps. Dashed line shows the location of the Yung-An Lineament (YAL) transpressional fault system. Circles show where BSRs are identified on the seismic profiles, and colors show BSR depths in meters below sea floor (mbsf).

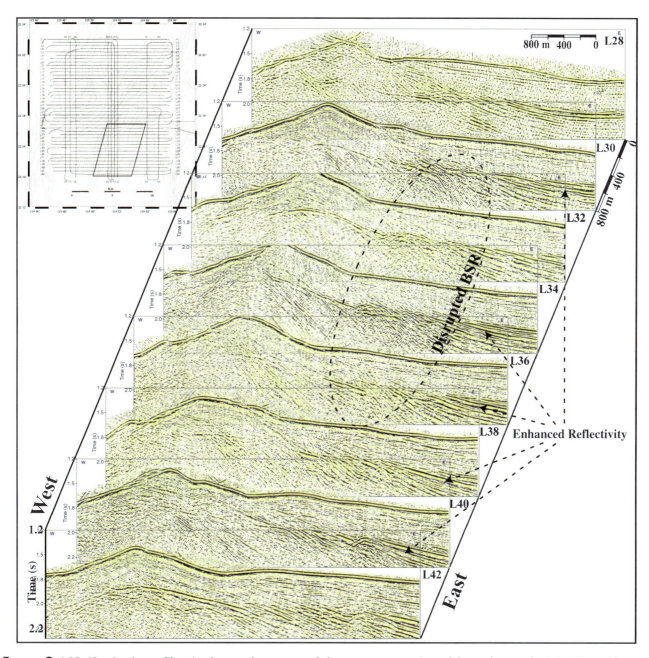

FIGURE 2. L28-42 seismic profiles, in the southern part of the survey area, viewed from the south. Seismic profiles are plotted in true amplitude, and the profile spacing is 800 m (2624 ft). Through time, the bottom-simulating reflector (BSR) migrates upward as the slope basin fills with new deposits, whereas the BSR deepens with respect to a fixed horizon in the uplifting anticline. With lower mobility, free-gas bubbles in the pore space are trapped beneath the BSR in the slope basin, resulting in bright reflectors but corresponding to the least amount of fluid- and gas-transport into the hydrate zone.

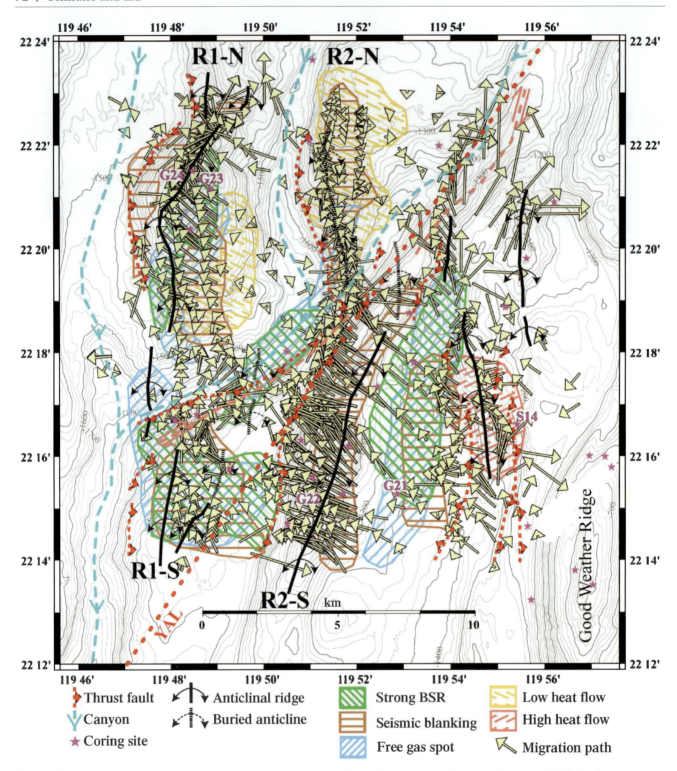

FIGURE 3. Estimated preferential fluid-migration paths across the bottom-simulating reflectors (BSRs) in the survey area. Arrows indicate the strike (azimuth) and dip of the strata at the BSR: arrow length is the sine of the difference between stratigraphic and BSR dips. Coring locations are indicated by stars. The most active fluid-migration zones correlate well with the high gas volumes sampled at the crest of the anticlinal ridges R1 and R2. Dashed line shows the location of the Yung-An Lineament (YAL).

20

Sava, D., and B. Hardage, 2009, Extended abstract—Rock-physics models
for gas-hydrate systems associated with unconsolidated marine sediments,
in T. Collett, A. Johnson, C. Knapp, and R. Boswell, eds., Natural gas
hydrates—Energy resource potential and associated geologic hazards: AAPG
Memoir 89, p. 73–75.

Extended Abstract—Rock-physics Models for Gas-hydrate Systems Associated with Unconsolidated Marine Sediments

Diana Sava and Bob Hardage
Bureau of Economic Geology, Jackson School of Geosciences, The University of Texas at Austin, Austin, Texas, U.S.A.

ABSTRACT

Rock-physics models are presented describing gas-hydrate systems associated with unconsolidated marine sediments. The goals are to predict gas-hydrate concentration from seismic attributes, such as compressional- and shear-wave (P and S wave) velocities, and to analyze compressional-wave (PP) and converted shear-wave (PS) reflectivity at the base of hydrate stability zones. Elastic properties of gas-hydrate systems depend on elastic properties of the host sediments, elastic properties of gas hydrates, concentration of hydrates in the sediments, and the geometrical details of hydrate morphology within the host sediments. We consider various scenarios for hydrate occurrence, including load-bearing gas hydrate, pore-filling gas hydrate, and two different thin-layered models of gas hydrate intercalated with unconsolidated sediments. We show that the geometrical details of how gas hydrates are distributed within sediments have a significant impact on relationships between gas-hydrate concentration and seismic attributes. Therefore, to accurately estimate gas-hydrate concentrations from seismic data, we need to understand how hydrates are formed and distributed within marine sediments. The modeling results for thin-layered hydrated morphologies show significant S-wave anisotropy, which may be used to infer gas-hydrate distributions and concentrations in alternating thin layers of hydrate-bearing sediments if multicomponent seismic data are available.

We compare the theoretical predictions of the isotropic rock-physics models with published laboratory measurements on synthetic gas-hydrate formed in unconsolidated sands. We find good agreement between the rock-physics model of disseminated, load-bearing gas hydrate and laboratory measurements, which suggests that, in this case, gas hydrates may act as part of the mineral frame of the unconsolidated sediments.

Model A: disseminated, load-bearing

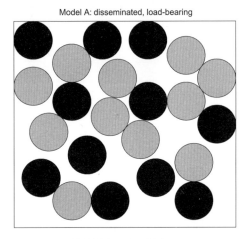

Model B: disseminated, nonload-bearing

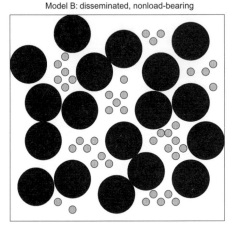

FIGURE 1. Graphical sketch of the four models of gas hydrate systems: load-bearing gas hydrates (model A); pore-filling gas hydrates (model B); thin layers of pure gas hydrate intercalated with unconsolidated sediments (model C); thin layers of disseminated, load-bearing gas hydrates intercalated with unconsolidated sediments (model D). Hydrates are represented in gray and sediment in black. Figure used with permission of The Leading Edge.

Model C: layered, solid phase

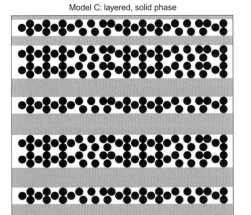

Model D: layered, disseminated phase

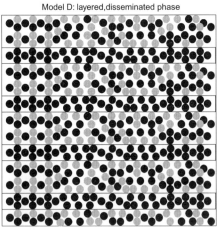

● Hydrate   ● Sediment

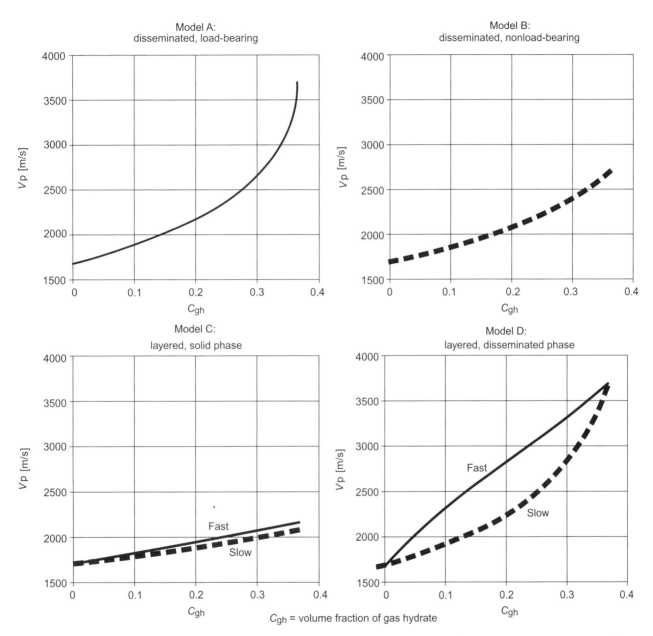

FIGURE 2. Primary-wave velocity as a function of the volumetric fraction of gas hydrate (c_{gh}) in pure quartz sediments for the four rock-physics models considered. Model A: Load-bearing gas hydrates disseminated in the whole volume of sediments; model B: pore-filling gas hydrates disseminated in the whole volume of sediments; model C: layers of pure gas hydrates producing slow P waves (dotted line) and fast P waves (solid line); model D: layers of disseminated, load-bearing gas hydrates producing slow P waves (dotted line) and fast P waves (solid line). 1000 m = 3281 ft. Figure used with permission of The Leading Edge.

Lee, M. W., T. S. Collett, and W. F. Agena, 2009, Extended abstract—Integration of vertical seismic, surface seismic, and well-log data at the Mallik 2L-38 gas-hydrate research well, Mackenzie delta, Canada, in T. Collett, A. Johnson, C. Knapp, and R. Boswell, eds., Natural gas hydrates—Energy resource potential and associated geologic hazards: AAPG Memoir 89, p. 77–79.

21

Extended Abstract—Integration of Vertical Seismic, Surface Seismic, and Well-log Data at the Mallik 2L-38 Gas-hydrate Research Well, Mackenzie Delta, Canada

M. W. Lee, T. S. Collett, and W. F. Agena

U.S. Geological Survey, Denver, Colorado, U.S.A.

ABSTRACT

Vertical-seismic-profile (VSP) data acquired at the Mallik 2L-38 well, Mackenzie delta, Northwest Territories, Canada, were analyzed and combined with surface seismic and downhole well-log data to (1) estimate gas-hydrate concentration around the well and (2) characterize the arctic gas-hydrate accumulations using different scale lengths ranging from 0.3 (sonic log) to 60 m (surface seismic) (0.9 to 197 ft). The interval compressional (P-)wave velocities derived from VSP data are somewhat slower than those from the well-log data. Furthermore, the shear (S-)wave velocities derived from VSP data within the depth interval of 600–900 m (1968–2953 ft) are about 20% slower than the sonic-log-derived velocity, implying seismic anisotropy. The spectral ratio of downgoing waves indicates that the P-wave attenuation quality factor of non–gas-hydrate-bearing sediments is about 65, whereas that of gas-hydrate-bearing sediments is about 170. The seismically determined thickness of gas-hydrate-bearing sediments inside a 2.3 × 2.4-km (1.42 × 1.49-mi) area surrounding the Mallik 2L-38 well is about 212 m (695 ft). Porosity obtained from well-log data averages 30%. The average gas-hydrate concentration estimated from the surface seismic data is about 43% of the pore space, and a cubic meter (35 cubic feet) of gas hydrate is 164 m³ (5792 ft³) of free gas. Therefore, the estimated gas content present in the gas-hydrate-bearing sediments is equivalent to 4.5×10^9 m³/km² (4.1×10^{11} ft³/mi²) of gas at the standard conditions (0°C and 1 atmosphere).

DOI:10.1306/13201162M893359

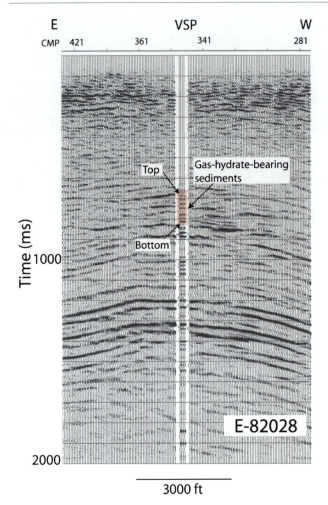

FIGURE 1. Surface seismic profile data (E-82028) with the insertion of a corridor stack of offset vertical-seismic-profile (VSP) data at the Mallik 2L-38 well, Mackenzie delta, Northwest Territories, Canada. The top of the gas-hydrate-bearing sediments is near 660 ms, and the bottom of the gas-hydrate-bearing sediments is near 820 ms (two-way traveltimes). Note that seismic amplitudes are much higher near the well. CMP = common midpoint.

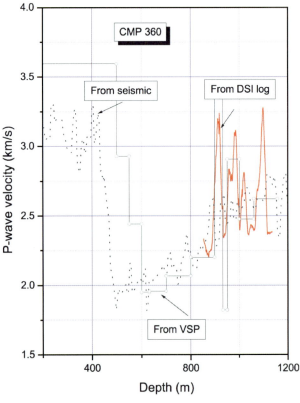

FIGURE 2. Result of a one-dimensional velocity inversion for common midpoint (CMP) 360 of E-82028 with vertical-seismic-profile (VSP)-derived and dipole shear sonic imager (DSI) well-log velocities. The DSI sonic log velocity is averaged across more than 10 m (33 ft). The dominant frequency of the surface seismic data is about 30 Hz. Using an interval velocity of 2600 m/s (8530 ft/s), the dominant wavelength is about 90 m (295 ft).

Gas Hydrate

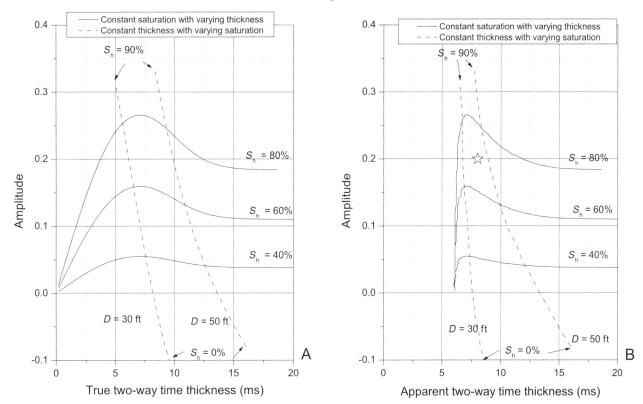

FIGURE 3. A) Graph showing the relationship between true two-way time of reservoir thickness, gas-hydrate saturation, and amplitude using a 55-Hz Ricker wavelet. Solid lines represent the relationship between amplitude, thickness, and three gas-hydrate saturations. Dotted lines represent the relationship between amplitude, gas-hydrate saturations, and thicknesses. B) The same as panel A except that the x axis is the apparent two-way time of reservoir thickness. The star represents an arbitrary data point to illustrate the interpretation procedure shown in the main text. The two-way apparent time thickness of the star is about 8 ms and the normalized amplitude is 0.2, which yields a gas-hydrate saturation of about 70% and thickness of 13.5 m (44.2 ft). S_h = gas-hydrate saturation; D = thickness; 1 m = 3.3 ft.

23

Inks, T. L., M. W. Lee, W. F. Agena, D. J. Taylor, T. S. Collett, M. V. Zyrianova, and R. B. Hunter, 2009, Extended abstract—Seismic prospecting for gas-hydrate and associated free-gas prospects in the Milne Point area of northern Alaska, *in* T. Collett, A. Johnson, C. Knapp, and R. Boswell, eds., Natural gas hydrates—Energy resource potential and associated geologic hazards: AAPG Memoir 89, p. 85–89.

Extended Abstract—Seismic Prospecting for Gas-hydrate and Associated Free-gas Prospects in the Milne Point Area of Northern Alaska

T. L. Inks

Interpretation Services, Inc., Denver, Colorado, U.S.A.

M. W. Lee, W. F. Agena, D. J. Taylor, T. S. Collett, and M. V. Zyrianova

U.S. Geological Survey, Denver, Colorado, U.S.A.

R. B. Hunter

Arctic Slope Regional Corporation Energy Services, Anchorage, Alaska, U.S.A.

ABSTRACT

The 1995 U.S. Geological Survey assessment of the in-place natural-gas-hydrate resources of the United States suggested that permafrost-associated gas hydrates on the Alaska North Slope may contain as much as 590 tcf (16.7 tcm) of in-place gas. However, this gas hydrate assessment did not specifically identify or characterize the nature of individual gas-hydrate accumulations or prospects. Detailed analysis and interpretation of two-dimensional (2-D) and three-dimensional (3-D) seismic data, along with seismic modeling and correlation with specially processed downhole well-log data, have led to the development of a viable method for identifying subpermafrost gas-hydrate prospects within the gas-hydrate stability zone (GHSZ) and associated sub–gas-hydrate free-gas prospects in the Milne Point area of northern Alaska. A total of 14 prospects exhibiting zones of higher amplitude within the GHSZ but below permafrost were expected to be gas-hydrate-bearing based on seismic modeling. The identified gas-hydrate prospects are typically fault bounded and are identified primarily by their acoustic properties. In most cases, areas that are currently structurally high within the imaged fault blocks can be shown to have acoustic properties that correspond to

Copyright ©2009 by The American Association of Petroleum Geologists.
DOI:10.1306/13201164M893360

higher concentrations of gas hydrate. This structural relationship is similar to conventional gas prospects, suggesting a probable free-gas origin. It is also clear that some of these fault-controlled structures are not fully charged with gas or gas hydrates, because down-dip limits to the mapped acoustic anomalies exist. Seismic data, in conjunction with geophysical modeling results from a related study, were used to further characterize the thickness and concentration of gas-hydrate occurrences within the delineated prospects. A Monte-Carlo-style statistical analysis of the seismic and well-log-derived reservoir data indicates that the gas-hydrate prospects in the Milne Point area may hold about 668.2 bcf (18.9 bcm) of gas.

Of the Milne Point gas-hydrate prospects, the Mount Elbert prospect is the best-defined prospect identified in the study and is unique because of its aerial extent, structural compartmentalization, potential for multiple pay zones, and its proximity to existing infrastructures. A stratigraphic test well within the Mount Elbert prospect was drilled in February 2007 and was found to have thick C and D unit gas-hydrate reservoirs (as defined by Collett [1993, 2002] in their study of Eileen hydrate accumulations) as predicted in this study.

REFERENCE CITED

Collett, T. S., 1993, Natural gas hydrates of the Prudhoe Bay and Kuparuk River area, North Slope, Alaska: AAPG Bulletin, v. 77, no. 5, p. 793–812.

Collett, T. S., 2002, Energy resource potential of natural gas hydrates; Unconventional petroleum systems: AAPG Bulletin, v. 86, no. 11, p. 1971–1992.

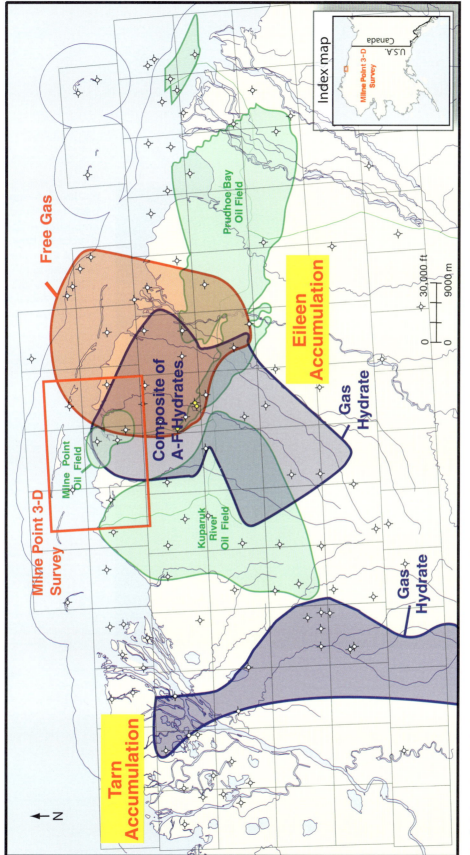

FIGURE 1. Milne Point 3-D study area and areas of identified gas-hydrate occurrence based on subsurface well-log studies. The Milne Point study area is in the northernmost part of the Eileen gas-hydrate accumulation. The study area is adjacent to the prolific Kuparuk River and Prudhoe Bay oil fields, and overlies the Milne Point oil field.

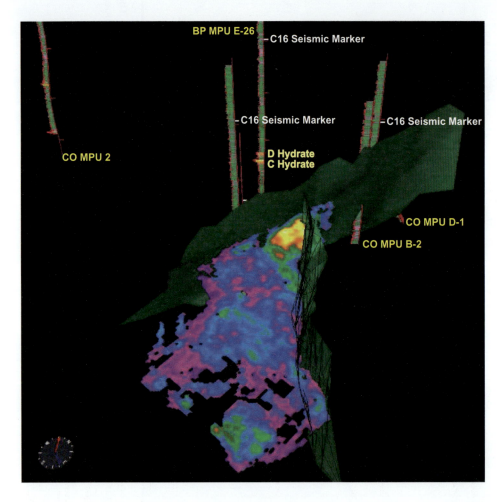

FIGURE 2. Mount Elbert prospect, fault-bounded and amplitude-draped over the structure, shows the relationship of the acoustic anomaly to bounding faults and the BP Milne Point Unit (MPU) E-26 well with thin unit C gas-hydrate reservoirs in 3-D.

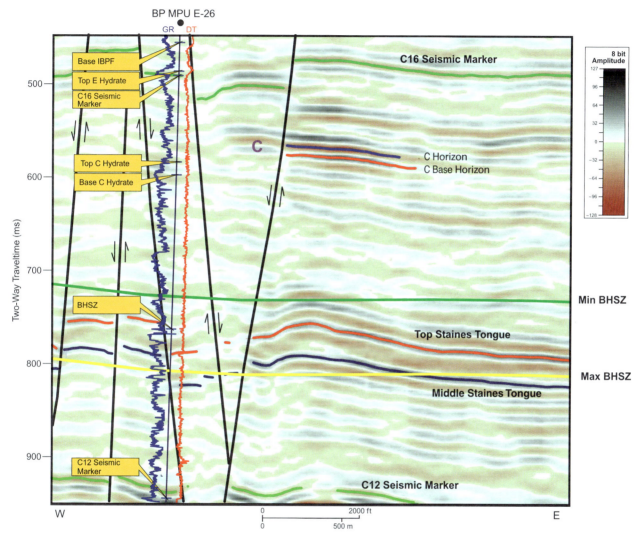

FIGURE 3. The seismically inferred Mount Elbert unit C gas-hydrate occurrence is imaged as a high-amplitude event on this line extracted from the Milne Point seismic volume at a depth of about 560 ms (two-way traveltime). The tie to the British Petroleum (BP) Milne Point Unit (MPU) E-26 well is shown. The gamma ray curve (GR) is shown on the left in blue, and the sonic curve (DT) is shown on the right in red. See Collett (1993, 2002) for definitions of A–F hydrates for the depth Slope Alaska Eileen hydrate accumulations.

Lorenson, T. D., T. S. Collett, and M. J. Whiticar, 2009, Extended abstract—
Hydrocarbon gas composition and origin of gas hydrate from the Alaska
North Slope, in T. Collett, A. Johnson, C. Knapp, and R. Boswell, eds.,
Natural gas hydrates—Energy resource potential and associated geologic
hazards: AAPG Memoir 89, p. 91–93.

24

Extended Abstract—Hydrocarbon Gas Composition and Origin of Gas Hydrate from the Alaska North Slope

Thomas D. Lorenson

U.S. Geological Survey, Menlo Park, California, U.S.A.

Timothy S. Collett

U.S. Geological Survey, Denver Colorado, U.S.A.

Michael J. Whiticar

School of Earth and Ocean Sciences, University of Victoria, Victoria, British Columbia, Canada

ABSTRACT

Hydrocarbon gas composition and isotopic composition of methane were analyzed from drill-cutting samples obtained from industry oil wells penetrating the Eileen and Tarn gas-hydrate deposits. These gas-hydrate deposits overlie the Prudhoe Bay and Kuparuk River oil fields and are restricted to the updip part of a series of nearshore deltaic sandstone reservoirs in the lower Tertiary (Eocene) Mikkelsen Tongue of the Canning Formation and the Tertiary Staines Tongue of the Sagavanirktok Formation, respectively. The Eileen gas hydrates occur in six laterally continuous Tertiary sandstone units, with individual occurrences in the range of 3–31 m (10–102 ft) thick. A significant accumulation of free hydrocarbon gas occurs downdip to the northeast, roughly at 700 m (2296 ft), where the sandstone units cross the structure I gas-hydrate stability zone. In the Tarn oil field, 50 km (31 mi) to the southwest, gas hydrate occurs as a 60–70-m (197–230-ft) thick zone within the Tertiary Ugnu and West Sak sandstones at about 230–300-m (754–984-ft) depth. The base of methane-hydrate stability is about 530 m (1738 ft).

Gas, mainly methane, that formed the gas hydrate either directly converted to gas hydrate or was first concentrated in existing conventional traps and later converted to gas hydrate in response to climate cooling or changes in surface conditions. Both the Eileen and Tarn gas hydrates are thought to contain a mixture of deep-sourced thermogenic gas, microbially modified thermogenic gas resulting from the anaerobic biodegradation of oil, and shallow microbial gas (methane carbon isotopic composition ranges from −54 to −46‰ in the gas-hydrate stability zone). The average methane isotopic composition found in the gas-hydrate stability zone is characteristic of methane sourced from biodegraded oils via the carbonate reduction pathway (−45 to −55‰). Microbial gas is present above the gas-hydrate deposits (methane carbon isotopic composition ranges from −74 to −56‰).

Thermogenic gases likely come either from existing oil and gas accumulations or from source rocks within the oil- and gas-generating windows that have migrated updip and or upfault and formed gas hydrate. The microbially modified thermogenic gases likely have a large source contribution from biodegraded oil or gas from the underlying oil fields, as evidenced by the carbon isotopic composition of methane, ethane, propane, and carbon dioxide; the dry molecular composition; and high iC_4/nC_4 ratios. The biodegradation of oil or condensate may result in producing ethane and propane suggested by the very light isotopic composition of these gases. The isotopic compositions of ethane ($\delta^{13}C_2$ −44 to −53‰) and propane ($\delta^{13}C_3$ −47 to −36‰) are very similar to methane ($\delta^{13}C_1$ −51 to −39‰). The West Sak oils, a likely source of gas for gas hydrate, are classified as moderately to heavily biodegraded. This pool of biodegraded oil underlies the Eileen gas-hydrate deposits and is about 50 km (31 mi) northeast of the Tarn well. Some part of the gas forming the Tarn gas-hydrate accumulation may have migrated updip from the large accumulation of biodegraded oil in the West Sak and Ugnu sandstones. The Eileen gas hydrate accumulation likely has a similar gas component of the same biodegraded oil, because it is located directly over the West Sak and Ugnu sand oil reservoir, both of which are cut by the Eileen fault, which is known to act as a conduit for gas seeping all the way to the surface.

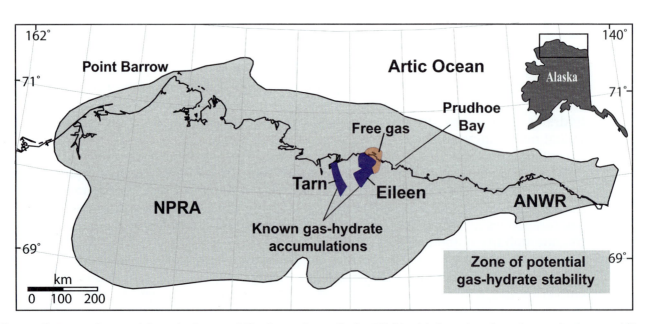

FIGURE 1. Zone of potential gas-hydrate stability in northern Alaska. Within this broad region, the gas-hydrate stability field exists both onshore and offshore and extends eastward into the Mackenzie delta region of Canada. The Eileen and Tarn gas-hydrate accumulations are shown in dark blue, and free gas underlying the Eileen gas-hydrate accumulation is shown in pink. NPRA refers to the National Petroleum Reserve, Alaska, and ANWR refers to the Arctic National Wildlife Refuge.

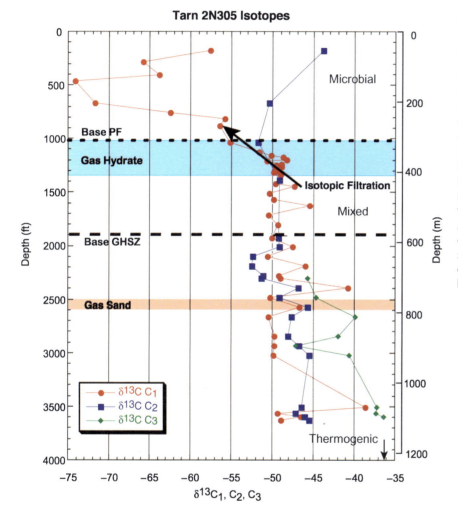

FIGURE 2. Profiles of methane, ethane, and propane carbon isotopic composition with depth from the Conoco-Phillips Tarn 2N305 well located in the southern section of the Tarn gas hydrate accumulation. The base of permafrost (base PF) and the base of the gas-hydrate stability zone (base GHSZ) are shown. Gas sources (microbial, mixed, thermogenic) and gas-hydrate and gas-sand occurrences are indicated. Note the isotopic filtration of methane that occurs between the bottom and the top of the gas hydrate zone. Note the light carbon isotopic composition of ethane and propane suggesting that biodegradation and/or evaporative fractionation of gases may be occurring.

25

Sigal, R. F., C. Rai, C. Sondergeld, B. Spears, W. J. Ebanks Jr., W. D. Zogg, N. Emery, G. McCardle, R. Schweizer, W. G. McLeod, and J. Van Eerde, 2009, Extended abstract—Characterization of a sediment core from potential gas-hydrate-bearing reservoirs in the Sagavanirktok, Prince Creek, and Schrader Bluff formations of Alaska's North Slope: Part 1—Project summary and geological description of core, in T. Collett, A. Johnson, C. Knapp, and R. Boswell, eds., Natural gas hydrates—Energy resource potential and associated geologic hazards: AAPG Memoir 89, p. 95–97.

Extended Abstract—Characterization of a Sediment Core from Potential Gas-hydrate-bearing Reservoirs in the Sagavanirktok, Prince Creek, and Schrader Bluff Formations of Alaska's North Slope: Part 1— Project Summary and Geological Description of Core*

R. F. Sigal, C. Rai, C. Sondergeld, and B. Spears
Mewbourne School of Petroleum and Geological Engineering, University of Oklahoma, Norman, Oklahoma, U.S.A.

W. J. Ebanks Jr.
Consultant, College Station, Texas, U.S.A.

W. D. Zogg[1], N. Emery, G. McCardle, and R. Schweizer
PTS Labs, Houston, Texas, U.S.A.

W. G. McLeod
Lone Wolf Oilfield Consulting, Calgary, Alberta, Canada

J. Van Eerde
Consultant, Calgary, Alberta, Canada

[1]*Present address*: Marathon Oil Corp., Houston, Texas, U.S.A.
Editor's note: This report is part of a five-report series (included in AAPG Memoir 89) on the geologic, petrophysical, and geophysical analysis of a sediment core recovered from the Hot Ice 1 gas-hydrate research well drilled in northern Alaska during 2003–2004. Each of these reports deals with specific topical observations and/or core measurements, including (Part 1) Project Summary and Geological Description of Core; (Part 2) Porosity, Permeability, Grain Density, and Bulk Modulus Core Studies; (Part 3) Electrical Resistivity Core Studies; (Part 4) Nuclear Magnetic Resonance (NMR) Core Studies; and (Part 5) Acoustic Velocity Core Studies.

ABSTRACT

The Anadarko Hot Ice 1 well was cored with a continuous coring system that recovered a 3.25-in. (8.25-cm) diameter core from 107 to 2300 ft (33 to 701 m) below the surface. Core recovery was 94.5%. The coring proceeded in two phases, separated by a summer of no activity. In phase I, the core was acquired from 107 to 1400 ft (33 to 427 m). With the exception of the bottom 140 ft (43 m), the phase I core was from the permafrost interval. The base of the permafrost interval was identified at a depth of 1260 ft. The core from the permafrost zone was recovered in a frozen state. The recovered core was described and characterized at the drill site.

The observations and measurements made at the drill site along with the subsequent analysis are described in five individual reports published in this Memoir. This Project Summary and Geological Description of Core report contains a detailed project review and description of the Hot Ice project and highlights from each of the five technical reports included in this volume.

The Hot Ice core penetrated 7 ft (2 m) of surface gravel and then entered the Sagavanirktok Formation. The Hot Ice well cored through the early Tertiary age sediments of the Sagavanirktok Formation, the Tertiary and upper Cretaceous Prince Creek Formation (which includes the informally named Ugnu sands), and Schrader Bluff Formation (which includes the informally named West Sak sands); the core ended in 42 ft (13 m) of what appears to be a marine section of fine-grained sediment. The sediments above the marine section were identified from the visual core description as deposited in environments that ranged from fluvial to marginal marine and upper deltaic. Alternating sandstone, mudstone, conglomerate, and coal formations form sequences that indicate an overall progradation and shallowing of environments of deposition with time.

The core from the first phase of the project, phase I, after the melting of the pore ice, was completely unconsolidated. Core recovery through the bottom of the mudstone in which phase I drilling was terminated was 96.3%. The core recovered from this interval contains 12.4% conglomerate, 43.50% sandstone, 37.8% mudstone, and 6.3% coal.

The second phase of the coring program, phase II, recovered core from 1403 ft (428 m) subsurface to 2300 ft (701 m). The bottom of the hole was estimated to be below the base of the hydrate stability zone at the Hot Ice location. The first 59 ft (18 m) of the core from phase II consisted of the marine sediments in which the phase I drilling program ended. The remainder of the core was composed of sediments from the Schrader Bluff Formation (West Sak section).

The sediments cored during phase II are not as variable in character as those encountered during phase I and are more fine grained. Layers of shell fragments and whole bivalve shells are common. The sandstones and mudstones in this interval form upward-coarsening sequences and have gradational contacts with the sediments above and below. These features suggest that the sediments in this interval may have been deposited in a shallow, marine shelf environment.

Unlike the phase I core, the phase II core contains no conglomerate or coal. The core recovered below the marine sand in which phase I terminated contains 68% mudstone, 26.3% sandstone, and 5.7% siltstone. In general, the sandstones were finer grained than those of the Ugnu sands.

Gas hydrate was not encountered at the site of the Hot Ice 1 well; however, the recovered cores and associated laboratory core observations and analysis provide a critical data set for understanding the nature of the sediments that have been shown to be gas-hydrate-bearing in nearby wells.

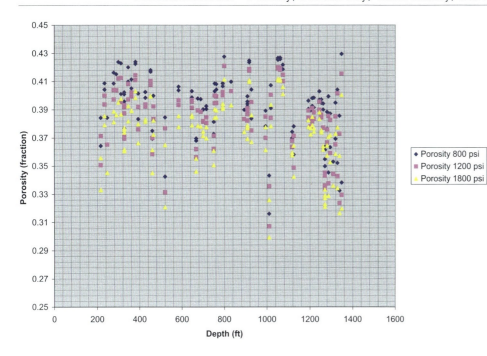

FIGURE 1. Porosity distribution for sands recovered during phase I. Porosity is given at three confining pressures: 800, 1200, and 1800 psi (5.5, 8.3, and 12.4 MPa). The porosity was computed from the measured pore volume divided by the pore plus grain volume. The zone below 1250 ft (381 m) may represent a lower porosity zone. 1 ft = 0.3 m.

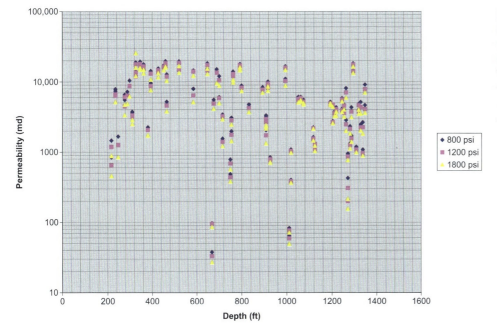

FIGURE 2. Permeability distribution for sands recovered in phase I. The permeability has been Klinkenberg corrected. The permeability is given for three confining pressures: 800, 1200, and 1800 psi (5.5, 8.3, and 12.4 MPa). 1 ft = 0.3 m.

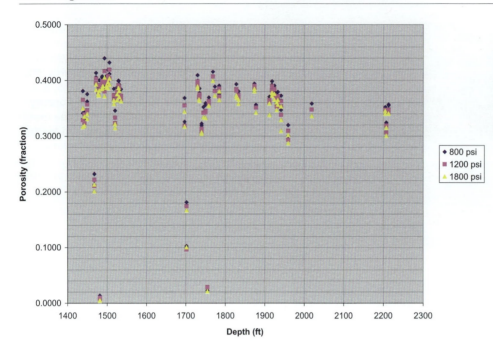

FIGURE 3. Porosity distribution for sands recovered during phase II. Porosity is given at three confining pressures: 800, 1200, and 1800 psi (5.5, 8.3, and 12.4 MPa). The porosity was computed from the measured pore volume divided by the pore plus grain volume. 1 ft = 0.3 m.

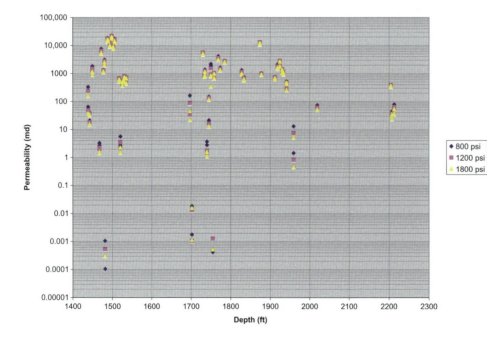

FIGURE 4. Distribution with depth of the measured Klinkenberg-corrected permeability values for the phase II recovered samples. The values less than 0.1 md are located in the anomalous tight zones. The permeability is given for three confining pressures: 800, 1200, and 1800 psi (5.5, 8.3, and 12.4 MPa). 1 ft = 0.3 m.

27

Sigal, R. F., C. Rai, C. Sondergeld, B. Spears, W. J. Ebanks Jr., W. D. Zogg, N. Emery, G. McCardle, R. Schweizer, W. G. McLeod, and J. Van Eerde, 2009, Extended abstract—Characterization of a sediment core from potential gas-hydrate-bearing reservoirs in the Sagavanirktok, Prince Creek, and Schrader Bluff Formations of Alaska's North Slope: Part 3—Electrical resistivity core studies, *in* T. Collett, A. Johnson, C. Knapp, and R. Boswell, eds., Natural gas hydrates—Energy resource potential and associated geologic hazards: AAPG Memoir 89, p. 103–106.

*Extended Abstract—Characterization of a Sediment Core from Potential Gas-hydrate-bearing Reservoirs in the Sagavanirktok, Prince Creek, and Schrader Bluff Formations of Alaska's North Slope: Part 3— Electrical Resistivity Core Studies**

R. F. Sigal, C. Rai, C. Sondergeld, and B. Spears
Mewbourne School of Petroleum and Geological Engineering, University of Oklahoma, Norman, Oklahoma, U.S.A.

W. J. Ebanks Jr.
Consultant, College Station, Texas, U.S.A.

W. D. Zogg[1], N. Emery, G. McCardle, and R. Schweizer
PTS Labs, Houston, Texas, U.S.A.

W. G. McLeod
Lone Wolf Oilfield Consulting, Calgary, Alberta, Canada

J. Van Eerde
Consultant, Calgary, Alberta, Canada

[1]*Present address*: Marathon Oil Corp., Houston, Texas, U.S.A.
**Editor's note*: This report is part of a five-report series (included in AAPG Memoir 89) on the geologic, petrophysical, and geophysical analysis of a sediment core recovered from the Hot Ice 1 gas hydrate research well drilled in northern Alaska during 2003–2004. Each of these reports deals with specific topical observations and/or core measurements, including (Part 1) Project Summary and Geological Description of Core; (Part 2) Porosity, Permeability, Grain Density, and Bulk Modulus Core Studies; (Part 3) Electrical Resistivity Core Studies; (Part 4) Nuclear Magnetic Resonance (NMR) Core Studies; and (Part 5) Acoustic Velocity Core Studies.

ABSTRACT

The Anadarko Hot Ice 1 well was cored as part of a project to study the occurrence of gas hydrate on the North Slope of Alaska. The observations and measurements made at the drill site along with the subsequent core analysis are described in five individual reports published in this Memoir. This report deals with the electrical resistivity measurements made on the recovered core.

The measurement and analysis of core electrical resistivities followed different procedures depending on the sample type. The sandstone samples recovered in phase II were from unfrozen sediments. These samples were processed after cleaning and drying and resaturating with brine. Porosity and formation factor were measured as a function of confining pressure. Shale samples recovered from this section of the core hole were measured at their recovered states. The phase I recovered sandstone samples came from the permafrost zone. They were first measured at below-freezing temperature conditions with the pore fluids contained on recovery.

The phase II recovered sandstones were saturated with a 3% KCl solution, and resistivity was measured at 20°C at confining stresses ranging from 600 to 2200 psi (4.1 to 15.2 MPa). The cementation factor extracted from the core resistivity measurements had a median value of 1.94 at 800 psi (5.5 MPa) and 1.87 at 1800 psi (12.3 MPa). Two sand-core samples were also measured as recovered to estimate the salinity of the pore water in the phase II recovered cores. These estimates were 5400 and 13,000 ppm.

Two shale samples from the phase II recovered cores were measured in the recovered state. Within this study, both horizontal and vertical resistivities were measured. The horizontal sample measurements had resistivities of 3.26 and 3.25 ohm m, respectively. The corresponding vertical resistivity values done on companion plugs were 6.02 and 7.57 ohm m.

The resistivity analysis of the phase I recovered samples required assumptions and approximations because neither the percentage of unfrozen brine nor its salinity was known. The basic approximation made was that, for a plug containing both ice and unfrozen brine, the unfrozen brine had a salinity that is the lowest value needed to keep it from freezing at that temperature. We also assumed that the frozen samples satisfied Archie's law with a cementation factor of 2. These assumptions, along with the resistivity measurements, were used to estimate the unfrozen fluid porosity and the salinity before freezing of the pore fluid. With these assumptions, the estimated salinity of the brine in the pore space before the formation of permafrost had a median salinity of 7100 ppm. The percent of the pore space filled with brine is a function of temperature. The measurements ranged from 5% of the pore space unfrozen at −7.5°C (18.5°F) to 55% unfrozen at −1.5°C (29.3°F).

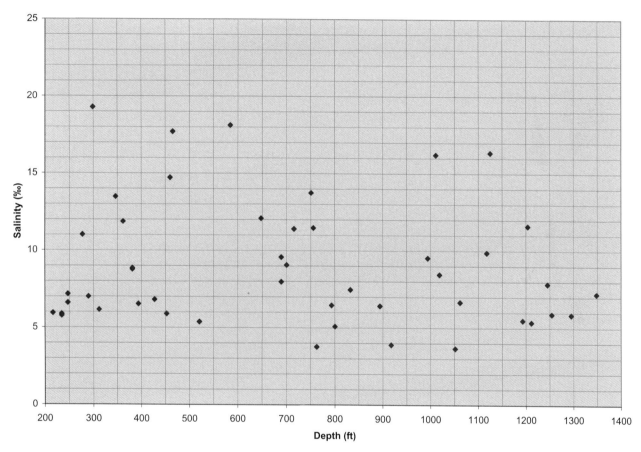

FIGURE 1. The estimated salinity of the brine in the pore space before the formation of permafrost and the partial freezing of the pore water in the sample. 1 ft = 0.3 m.

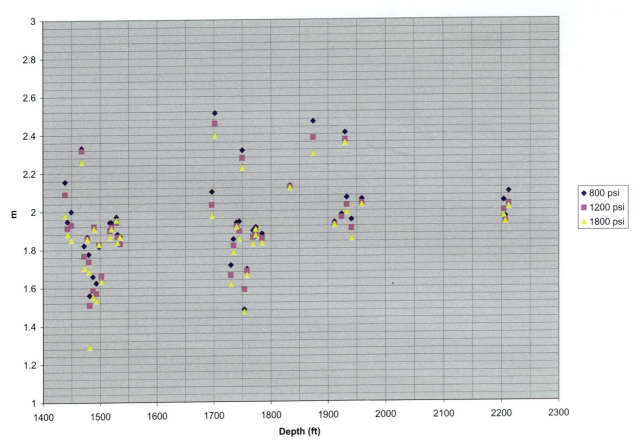

FIGURE 2. Archie's cementation coefficient *m* as a function of depth for three confining pressures. 100 psi = 0.7 MPa.

Winters, W. J., W. F. Waite, and D. H. Mason, 2009, Extended abstract—Effects of methane hydrate on the physical properties of sediments, *in* T. Collett, A. Johnson, C. Knapp, and R. Boswell, eds., Natural gas hydrates—Energy resource potential and associated geologic hazards: AAPG Memoir 89, p. 127–128.

33

Extended Abstract—Effects of Methane Hydrate on the Physical Properties of Sediments

William J. Winters, William F. Waite, and David H. Mason
U.S. Geological Survey, Woods Hole, Massachusetts, U.S.A.

ABSTRACT

Grain size, pore content, and arrangement of pore constituents have a profound effect on the acoustic and strength properties of sediments. We tested specimens containing gas hydrate, methane, and water in the pore space of coarse- and fine-grained sediments to simulate the marine environment and of frozen coarse-grained sediment to simulate permafrost conditions.

The measured compressional wave velocity (Vp) changes with the extent to which the pore material cements sediment grains. Hence, for equal effective stresses, V_p is lowest in gas-charged sediments, increases for water-saturated sediments, then increases significantly for hydrate-bearing sediments because of the sediment cementation provided by hydrate. Frozen sediment, effectively fully saturated and fully cemented sediment, exhibits the highest Vp.

Sediment strength follows the same pattern but also shows a strong dependence on sediment grain size. For consolidation stresses associated with the upper several hundred meters of subbottom depth, pore pressures decreased during shear in coarse-grained sediments containing gas hydrate, thereby increasing strength, whereas pore pressure in fine-grained sediments typically increased during shear, which decreased strength. The presence of free gas in pore space damped the pore pressure response during shear and reduced the strengthening effect of gas hydrate in sands.

DOI:10.1306/13201174M893365

FIGURE 1. Close-up view of a test specimen about to be raised into the main pressure vessel (visible at the top of the photograph). The test specimen (light brown cylinder), located in the central part of the photo, rests on an interchangeable internal load cell. A heat exchanger that imparts a unidirectional cooling front downward through the specimen rests atop the upper end cap and is fed through the large-diameter, vertical tubes at the front and rear of the specimen.

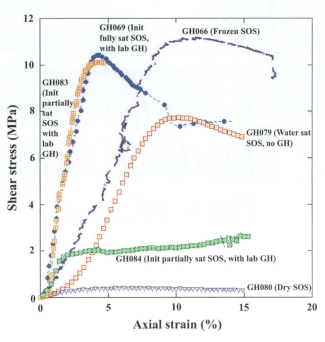

FIGURE 2. Shear-stress versus axial-strain plots for sieved Ottawa sand (SOS) specimens containing various pore-space materials. GH = gas hydrate; Init = initially; sat = saturated; 1 MPa = 145 psi.

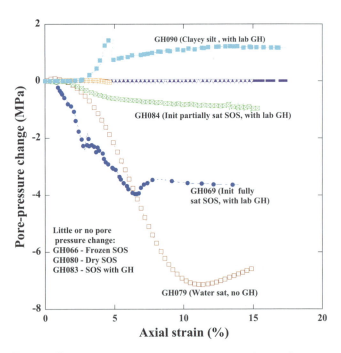

FIGURE 3. Pore-pressure change versus strain results for sieved Ottawa sand (SOS) (GH066, GH069, GH079, GH080, GH083, GH084) and clayey silt (GH090). The coarse-grained sediment containing gas hydrate and other pore fillings exhibited dilatant behavior, as illustrated by the generation of negative (related to the beginning of shear) pore pressure. Contrast this behavior with the positive pore-pressure response of GH090. GH083 contained a significant amount of gas hydrate that filled the sediment pores, resulting in little pore-pressure response. GH = gas hydrate; 1 MPa = 145 psi.

34

Jaiswal, N. J., A. Y. Dandekar, S. L. Patil, R. B. Hunter, and T. S. Collett, 2009,
Extended abstract—Relative permeability measurements of gas-water-
hydrate systems, *in* T. Collett, A. Johnson, C. Knapp, and R. Boswell, eds.,
Natural gas hydrates—Energy resource potential and associated geologic
hazards: AAPG Memoir 89, p. 129–130.

Extended Abstract—Relative Permeability Measurements of Gas-water-hydrate Systems

Namit J. Jaiswal

Shell Exploration and Production Company, Houston, Texas, U.S.A.

Abhijit Y. Dandekar and Shirish L. Patil

University of Alaska Fairbanks, Fairbanks, Alaska, U.S.A.

Robert B. Hunter

Arctic Slope Regional Corporation Energy Services, Anchorage, Alaska, U.S.A.

Timothy S. Collett

U.S. Geological Survey, Denver, Colorado, U.S.A.

ABSTRACT

A primary mechanism likely to control potential gas production from gas-hydrate-bearing porous media is the gas-water two-phase flow during dissociation. Gas-water relative-permeability functions within gas-hydrate systems are poorly understood, and direct measurements within gas-hydrate-bearing porous media are difficult. In this study, we developed a new method for measuring gas-water relative permeability for laboratory-synthesized gas hydrate within porous media. The new experimental design allows gas hydrate to form within a porous media and allows the measurement of effective permeability and relative permeability for different saturation values. The relative permeability to gas and water was determined by applying the Johnson-Bossler-Naumann method. Finally, effective permeability and relative permeability data of gas and water phases are reported for gas-hydrate-saturated consolidated Oklahoma 100-mesh sand and Alaska North Slope subsurface sediments.

The results show significant reduction in permeability at increased gas-hydrate saturations. The results also suggest that the relative permeability determined from the unsteady-state core floods is primarily affected by gas-hydrate saturations. Furthermore, effective as well as relative permeabilities vary by the nature of

DOI:10.1306/13201175M893366

gas-hydrate distribution for the same bulk saturation in different porous media. We believe that the experimental data obtained from this work will provide input data to reservoir modeling, fluid-flow modeling, and development of relative-permeability-estimation methods for hydrate production. However, considerable additional experimental and theoretical work remains to develop an analytical or generalized model to predict the relative permeability for gas-hydrate reservoir simulation.

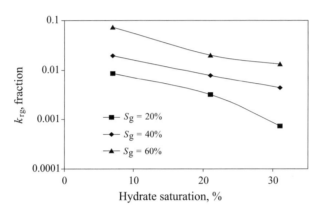

FIGURE 1. Gas relative permeability (k_{rg}) as a function of hydrate saturation for different isogas saturation values for Anadarko field sand samples.

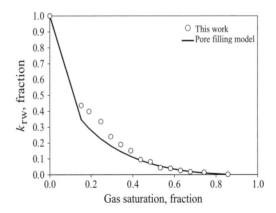

FIGURE 2. Comparison of measured (this work; Oklahoma 100-mesh sand sample) and estimated water relative (k_{rw}) permeability (pore filling model). The measured water relative permeability is in the presence of 10% hydrate saturation. The estimated values are from Spangenberg's (2001) pore filling model with exponent $n = 0.4$ for a hydrate saturation of 10%.

35

Minagawa, H., R. Ohmura, Y. Kamata, J. Nagao, T. Ebinuma, H. Narita, and Y. Masuda, 2009, Extended abstract—Water permeability of porous media containing methane hydrate as controlled by the methane-hydrate growth process, *in* T. Collett, A. Johnson, C. Knapp, and R. Boswell, eds., Natural gas hydrates—Energy resource potential and associated geologic hazards: AAPG Memoir 89, p. 131–133.

Extended Abstract—Water Permeability of Porous Media Containing Methane Hydrate as Controlled by the Methane-hydrate Growth Process

Hideki Minagawa, Ryo Ohmura, Yasushi Kamata, Jiro Nagao, Takao Ebinuma, and Hideo Narita
National Institute of Advanced Industrial Science and Technology (AIST), Sapporo, Japan

Yoshihiro Masuda
Department of Geosystem Engineering, School of Engineering, University of Tokyo, Tokyo, Japan

ABSTRACT

Methane hydrates (MHs) existing in the sea-floor sediment and permafrost are expected to be an unconventional methane resource. The methods of recovering methane gas from these hydrates, such as the depressurization method, the thermal stimulation method, and the inhibitor injection method, have already been proposed. In all methods, permeability to gas and water is an important property of the hydrate sediment for determining the recovery efficiency of methane gas. In this work, the relations between water permeability and hydrate saturation, which is the ratio of hydrate volume to the porosity volume, have been investigated. Furthermore, the relationship between water permeability and hydrate saturation with different hydrate formation processes are described.

The experimental apparatus measuring water permeability consists of a temperature-controlled core holder, water and gas inlet and outlet lines with pressure gauges, and a data-recording system. The core holder is combined with a hydrostatic triaxial pressurized system. The sample of sediments is set in a cylindrical rubber sleeve and pressurized from both axial and radial directions with a maximum pressure of 20 MPa (2901 psi).

We tested three different laboratory methods for the production of methane-hydrate-bearing sediments to determine the relationship between methane hydrate saturation (Sh) and permeability. The first method (Type 1) used a conventional approach, the connate (irreducible) water method. The second method (Type 2) developed by Chuvilin et al. (2003) is known as the gas diffusion method. The third method (Type 3) is based on a sediment-cementing process, performed on a mixture of fine-grained ice and sand.

From our experiments using each sample of sediment type under different production methods, we obtained data sets for apparent water permeability (AWP) versus Sh. Plotting the difference in AWP against Sh, we found that the AWP of sediments decreased with Sh. The AWPs measured by type 1 procedures were 20 times greater than the maximum value measured by type 2. These results indicate that the apparent permeability of each sample at the same Sh differs depending on the process used to grow the gas hydrate and to measure the apparent permeability. When we converted the relationship between water saturation (1 − Sh) and AWP to a logarithmic plot, the average slope for each production was obtained.

In theory, the apparent permeability K can be approximated by the following equation:

$$K = K_0(1 - \text{Sh})^N \qquad (1)$$

where K_0 is the apparent permeability at Sh = 0, and N is a constant. In type 1, the value of N is 2.6, in type 2 it is 9.8, and in type 3 it is 14.0.

Preliminary experiments employed x-ray computer tomography (CT) analysis for type 1 and 2 sediments that exhibited gas-saturation conditions after water saturation. The x-ray CT density image for type-1-generated gas-hydrate-bearing sediments was heterogeneous, whereas the scan of the sediment sample grown by the type 2 method was more homogeneous in nature. The heterogeneous density for the type 1 process sediment seems to be caused by a small channel of gas flow along the core sample that developed in the connate water formation process. In addition to the MH growth, the sediment structure would be affected by the permeability change with MH saturation.

REFERENCE CITED

Chuvilin, E. M., T. Ebinuma, Y. Kamata, T. S. Uchida, S. Takeya, J. Nagao, and H. Narita, 2003, Effects of temperature cycling on the phase transition of water in gas-saturated sediments: Canadian Journal of Physics, v. 81, no. 1–2, p. 343–350.

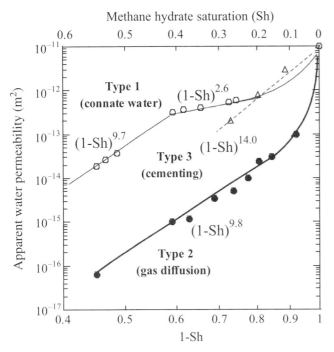

FIGURE 1. Plot of apparent water permeability and water saturation (1 − Sh). Sh = methane hydrate saturation.

Taylor, C. E., J. Lekse, and N. English, 2009, Extended abstract—Methane-hydrate laboratory and modeling research: Bridging the gap, *in* T. Collett, A. Johnson, C. Knapp, and R. Boswell, eds., Natural gas hydrates—Energy resource potential and associated geologic hazards: AAPG Memoir 89, p. 135–136.

36

Extended Abstract—Methane-hydrate Laboratory and Modeling Research: Bridging the Gap

Charles E. Taylor, Jonathan Lekse, and Niall English

U.S. Department of Energy, National Energy Technology Laboratory, Pittsburgh, Pennsylvania, U.S.A.

ABSTRACT

Methane hydrates are clathrates (crystalline solids whose building blocks consist of a gas molecule that stabilizes and is surrounded by a cage of water molecules) where methane is the guest molecule. Methane hydrates are stable and occur naturally in marine continental margins and permafrost-related sediments. At standard temperature and pressure (STP), one volume of saturated methane hydrates contains approximately 180 volumes of methane. Current estimates suggest that at least twice as much organic carbon is contained in methane hydrates exist as all other forms of fossil fuels combined. The methane hydrate deposits along the coast and in permafrost areas of the United States contain an estimated 320,000 tcf (9000 tcm) of methane. To tap into this vast resource, research is needed to understand the fundamental physical properties of hydrates.

This study is an introduction to the National Energy Technology Laboratory (NETL) hydrate facilities and capabilities. The NETL Methane Hydrate Research Group conducts research in four key areas: modeling, computation, thermodynamic properties, and kinetic properties. Our modeling focuses on flow simulation in reservoirs. Computational research models hydrate formation and dissociation. Thermodynamic properties research focuses on measurements of both synthetic and naturally occurring hydrates. Kinetic properties research measures the kinetic properties of methane hydrates (both synthetic and naturally occurring), including the physical properties of hydrates synthesized in one of the many view cells at NETL that range in volume from 1 mL to 15 L.

Copyright ©2009 by The American Association of Petroleum Geologists.
DOI:10.1306/13201177M893368

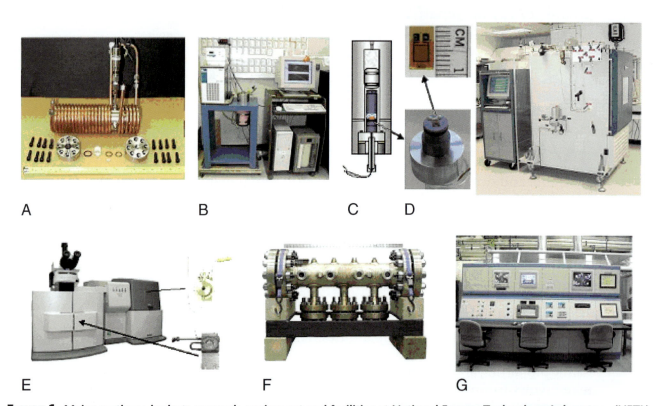

FIGURE 1. Major methane hydrate research equipment and facilities at National Energy Technology Laboratory (NETL). A) One of two 40-mL hydrate view cells. B) A 1-L hydrate cell. C) NETL-designed hydrate thermoconductivity cell. D) Environmental chamber and support equipment for the thermoconductivity cell. E) Raman spectrometer and the in-situ hydrate view cell. F) A 15-L hydrate view cell. G) Control housing for the 15-L hydrate view cell.

39

Zheng, L., H. Zhang, M. Zhang, P. Kerkar, and D. Mahajan, 2009, Extended abstract—Modeling methane-hydrate formation in marine sediments, *in* T. Collett, A. Johnson, C. Knapp, and R. Boswell, eds., Natural gas hydrates—Energy resource potential and associated geologic hazards: AAPG Memoir 89, p. 143–145.

Extended Abstract—Modeling Methane-hydrate Formation in Marine Sediments

Lili Zheng[1], Hui Zhang[2], and Mingyu Zhang[3]
Mechanical Engineering Department, State University of New York at Stony Brook, Stony Brook, New York, U.S.A.

Prasad Kerkar
Materials Science and Engineering Department, State University of New York at Stony Brook, Stony Brook, New York, U.S.A.

Devinder Mahajan
Energy Sciences and Technology Department, Brookhaven National Laboratory, Upton, New York, U.S.A. and Materials Science and Engineering Department, State University of New York at Stony Brook, Stony Brook, New York, U.S.A.

ABSTRACT

In this study, we review knowledge critical to simulating hydrate formation and dissociation in marine sediments. The advantages and disadvantages of existing numerical models are summarized. An advanced computational model (meshless particle-based model) is introduced to simulate fluid flow, heat and mass transfer, and hydrate formation at the pore level. In the model, the spatial distribution and uncertainty of porosity and variation of permeability with hydrate formation can be predicted. The model has been tested for different crystal growth kinetics. The pore-level simulation results may help us understand the quantity and distribution of methane hydrate within the confines of a pore space.

[1] *Present address*: School of Aerospace, Tsinghua University, China.
[2] *Present address*: Department of Engineering Physics, Tsinghua University, China.
[3] *Present address*: Institute of Applied Physics and Computational Mathematics, Beijing, China.

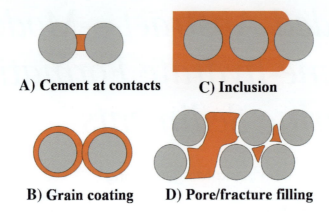

FIGURE 1. Pore-level methane-hydrate formation models. A) Hydrate cements at contact, B) hydrate growth around the grains, C) hydrate formation including grains, and D) hydrate as nodules and disseminated between pores and fractures.

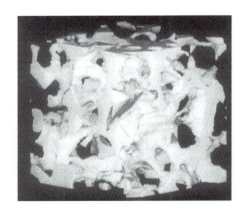

A)

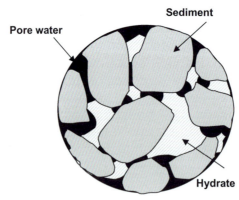

B)

FIGURE 2. A) View of the pore space of Fontainebleau sandstones, $275 \times 250\,\mu m$, obtained by computed microtomography (courtesy of Spanne et al., 1994). Reprinted with permission from Spanne et al., Physical Review Letters, v. 73, p. 2001, 1994. Copyright 1994 by the American Physical Society. B) schematic of hydrate formation in marine sediments.

A) no hydrate formation

B) hydrate formed at contacts

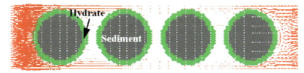

C) hydrate formed on particle surfaces

D) hydrate formed in the pore volume

FIGURE 3. Pore-level simulation showing water flowing around particles in a channel. The computational domain is 74 × 400 μm, the particle diameter is 50 μm, and the average velocity at the inlet is 0.02 m/s (0.06 ft/s). A) No hydrate formation, B) hydrate formed at contacts, C) hydrate formed on particle surfaces, and D) hydrate formed in the pore volume. The predicted fluid field changes dramatically because of hydrate formation, and particle space also varies during hydrate formation. This indicates that different geometric particles or pores can be simulated to study the interaction between hydrate formation and porous structure.

REFERENCE CITED

Spanne, P., J. F. Thovert, C. J. Jacquin, W. B. Lindquist, K. W. Jones, and P. M. Adler, 1994, Synchrotron computed microtomography of porous media: Topology and transports: Physical Review Letters, v. 73, p. 2001–2004.